Scattering, Natural Surfaces and Fractals

Scattering, Natural Surfaces and Fractals

Giorgio Franceschetti and Daniele Riccio

AMSTERDAM • BOSTON • HEIDELBERG • LONDON
NEW YORK • OXFORD • PARIS • SAN DIEGO
SAN FRANCISCO • SINGAPORE • SYDNEY • TOKYO

Academic Press is an imprint of Elsevier

Academic Press is an imprint of Elsevier
30 Corporate Drive, Suite 400, Burlington, MA 01803, USA
525 B Street, Suite 1900, San Diego, California 92101-4495, USA
84 Theobald's Road, London WC1X 8RR, UK

This book is printed on acid-free paper.

Copyright © 2007, Elsevier Inc. All rights reserved.

No part of this publication may be reproduced or transmitted in any form or by any means, electronic or mechanical, including photocopy, recording, or any information storage and retrieval system, without permission in writing from the publisher.

Permissions may be sought directly from Elsevier's Science & Technology Rights Department in Oxford, UK: phone: (+44) 1865 843830, fax: (+44) 1865 853333, E-mail: permissions@elsevier.com. You may also complete your request on-line via the Elsevier homepage (http://elsevier.com), by selecting "Support & Contact" then "Copyright and Permission" and then "Obtaining Permissions."

Library of Congress Cataloging-in-Publication Data
Franceschetti, Giorgio.
  Scattering, natural surfaces, and fractals/Giorgio Franceschetti and Daniele Riccio.
    p. cm.
  Includes bibliographical references and index.
  ISBN-13: 978-0-12-265655-2 (casebound : alk. paper)
  ISBN-10: 0-12-265655-5 (casebound : alk. paper)
    1. Electromagnetic waves–Scattering–Mathematical models.
  2. Surfaces (Physics)–Mathematical models.   I. Riccio, Daniele.
  II. Title.
  QC665.S3F73 2006
  530.14'1–dc22
                                                                    2006024471

British Library Cataloguing-in-Publication Data
A catalogue record for this book is available from the British Library.

ISBN 13: 978-0-12-265655-2
ISBN 10: 0-12-265655-5

For information on all Academic Press publications
visit our Web site at www.books.elsevier.com

Printed in the United States of America
06 07 08 09 10   9 8 7 6 5 4 3 2 1

Working together to grow
libraries in developing countries

www.elsevier.com | www.bookaid.org | www.sabre.org

ELSEVIER   BOOK AID International   Sabre Foundation

Dedication

To my wife, Giuliana
(Giorgio)

To my parents, Gennaro and Lucia
(Daniele)

Contents

Preface   xiii

1. The Scattering Problem   1

   1.1 Introduction and Chapter Outline   1
   1.2 The Scattering-Problem Definition   2
   1.3 Motivations   3
   1.4 Surface Models and Electromagnetic Methods   4
   1.5 Deterministic versus Stochastic Models for the Natural Surfaces   5
       1.5.1 Surface Deterministic Models   5
       1.5.2 Surface Stochastic Models   6
   1.6 Deterministic versus Stochastic Evaluation for the Scattered Field   9
       1.6.1 Scattered-Field Deterministic Descriptions   9
       1.6.2 Scattered-Field Stochastic Descriptions   10
   1.7 Analytic versus Numerical Evaluation of the Scattered Field   12
   1.8 Closed-Form Evaluation of the Electromagnetic Field Scattered from a Natural Surface   14
   1.9 Book Outline   18
   1.10 References and Further Readings   19

Contents

2. Surface Classical Models — 21

2.1 Introduction and Chapter Outline — 21
2.2 Fundamentals of Stochastic Processes — 22
    2.2.1 Stochastic Processes: Definition — 22
    2.2.2 Stochastic Processes: Relevant Averages — 23
    2.2.3 Stochastic Processes: a Relevant Property — 25
2.3 Spectral Characterization of Stochastic Processes — 26
2.4 Isotropic Surfaces — 30
2.5 Classical Models for Natural Surfaces: First-Order Stochastic Characterization — 33
2.6 Classical Models for Natural Surfaces: Second-Order Stochastic Characterization — 34
2.7 Physical Counterpart of Natural-Surfaces Classical Parameters — 40
    2.7.1 Standard Deviation — 41
    2.7.2 Correlation Length — 41
2.8 Surface Classical Models Selection for Electromagnetic Scattering — 44
2.9 References and Further Readings — 44
Appendix 2.A Surface Classical Models — 45
    2.A.1 Gaussian Autocorrelation — 46
    2.A.2 Exponential Autocorrelation — 47
    2.A.3 Intermediate Gaussian-Exponential Autocorrelation — 48
    2.A.4 Power-Law Autocorrelation — 50
    2.A.5 Multiscale Gaussian Autocorrelation — 51
    2.A.6 Multiscale Exponential Autocorrelation — 55
    2.A.7 Mixed Gaussian-Exponential Autocorrelation — 56

3. Surface Fractal Models — 61

3.1 Introduction and Chapter Outline — 61
3.2 Fundamentals of Fractal Sets — 63
    3.2.1 Hausdorff Measure — 64
    3.2.2 Fractal Dimension — 65
    3.2.3 Scaling Properties — 67
3.3 Mathematical versus Physical Fractal Sets — 68
3.4 Deterministic versus Stochastic Fractal Description of Natural Surfaces — 71

	3.5	Fractional Brownian Motion Process	72
		3.5.1 Mathematical fBm Processes	72
		3.5.2 Physical fBm Processes	91
	3.6.	Weierstrass-Mandelbrot Function	94
		3.6.1 Mathematical WM Functions	94
		3.6.2 Physical WM Functions	96
	3.7	Connection between fBm and WM Models	98
	3.8	A Chosen Reference Fractal Surface for the Scattering Problem	100
	3.9	Fractal-Surface Models and their Comparison with Classical Ones	101
		3.9.1 Classical Parameters for the fBm Process	103
		3.9.2 Classical Parameters for the WM Function	105
	3.10	References and Further Readings	106
		Appendix 3.A Generalized Functions	106
		Appendix 3.B Space-Frequency and Space-Scale Analysis of Nonstationary Signals	107
		3.B.1 Introduction	107
		3.B.2 fBm Wigner-Ville Spectrum	108
		3.B.3 fBm and Wavelet Approach	110

4. Analytic Formulations of Electromagnetic Scattering — 115

4.1	Introduction and Chapter Outline	115
4.2	Maxwell Equations	116
4.3	The Integral-Equation Method	118
4.4	Incident and Scattered-Field Coordinate-Reference Systems	125
4.5	The Kirchhoff Approximation	128
4.6	Physical-Optics Solution	132
4.7	Extended-Boundary-Condition Method	134
4.8	Small-Perturbation Method	138
4.9	References and Further Readings	140

5. Scattering from Weierstrass-Mandelbrot Surfaces: Physical-Optics Solution — 143

5.1	Introduction and Chapter Outline	143
5.2	Analytic Derivation of the Scattered Field	144

	5.3	Scattered-Field Structure	148
		5.3.1 Number of Modes Significantly Contributing to the Scattered Field	149
		5.3.2 Mode Directions of Propagation	153
		5.3.3 Modes Amplitude and Phase	155
	5.4	Limits of Validity	156
	5.5	Influence of Fractal and Electromagnetic Parameters over the Scattered Field	157
		5.5.1 The Role of the Fundamental-Tone Wavenumber	159
		5.5.2 The Role of the Tone Wave-Number Spacing Coefficient	162
		5.5.3 The Role of the Number of Tones	164
		5.5.4 The Role of the Overall Amplitude-Scaling Factor	166
		5.5.5 The Role of the Hurst Exponent	168
	5.6	Statistics of the Scattered Field	169
	5.7	References and Further Readings	170

6. Scattering from Fractional Brownian Surfaces: Physical-Optics Solution — 171

	6.1	Introduction and Chapter Outline	171
	6.2	Scattered Power-Density Evaluation	172
		6.2.1 Persistent fBm	175
		6.2.2 Antipersistent fBm	176
	6.3	Scattered Power Density	179
	6.4	Scattered Power Density: Special Cases	180
		6.4.1 Scattering in the Specular Direction	181
		6.4.2 Brownian Surfaces ($H=1/2$)	181
		6.4.3 Marginally Fractal Surfaces ($H \to 1$)	181
		6.4.4 Quasi-Smooth Surfaces ($kT \ll 1$)	182
	6.5	Backscattering Coefficient	182
	6.6	Validity Limits	184
	6.7	Influence of Fractal and Electromagnetic Parameters over the Scattered Field	186
		6.7.1 The Role of the Spectral Amplitude	188
		6.7.2 The Role of the Hurst Exponent	189
		6.7.3 The Role of the Electromagnetic Wavelength	191
	6.8	References and Further Readings	191

7. Scattering from Weierstrass-Mandelbrot Profiles: Extended-Boundary-Condition Method   193

    7.1   Introduction and Chapter Outline   193
    7.2   Profile Model   195
    7.3   Setup of the Extended-Boundary-Condition Method   196
           7.3.1   Incident Field   196
           7.3.2   Integral Equations   197
           7.3.3   Surface-Field Expansions for WM Profiles   201
    7.4   Surface-Fields Evaluation   204
    7.5   Fields Expansions   205
    7.6   EBCM Equations in Matrix Form   207
    7.7   Matrix-Equations Solution   209
    7.8   Matrices Organizations   210
    7.9   Scattering-Modes Superposition, Matrices Truncation, and Ill-Conditioning   213
    7.10   Influence of Fractal and Electromagnetic Parameters over the Scattered Field   215
           7.10.1   The Role of the Fundamental-Tone Wavenumber   219
           7.10.2   The Role of the Tone Wavenumber Spacing Coefficient   219
           7.10.3   The Role of the Number of Tones   223
           7.10.4   The Role of the Overall Amplitude-Scaling Factor   225
           7.10.5   The Role of the Hurst Exponent   227
    7.11   References and Further Readings   229
         Appendix 7.A Evaluation of the Dirichlet- and Neumann-Type Integrals   230
           7.A.1   Evaluation of the Dirichlet-Type Integral   231
           7.A.2   Evaluation of the Neumann-Type Integral   234

8. Scattering from Fractional Brownian Surfaces: Small-Perturbation Method   239

    8.1.   Introduction and Chapter Outline   239
    8.2.   Rationale of the SPM Solution   240
    8.3.   Extended Boundary Condition Method in the Transformed Domain   243
    8.4.   Set up the Small Perturbation Method   248

	8.5.	An Appropriate Coordinate System	251
	8.6.	Zero-order Solution	252
	8.7.	First-order Solution	256
	8.8.	Small Perturbation Method Limits of Validity	260
	8.9.	Influence of Fractal and Electromagnetic Parameters Over the Scattered Field	262
		8.9.1. The Role of the Spectral Amplitude	263
		8.9.2. The Role of the Hurst Exponent	264
		8.9.3. The Role of the Electromagnetic Wavelength	266
	8.10.	References and Further Readings	267

Appendix A: Mathematical Formulae   269

Appendix B: Glossary   273

Appendix C: References   277

Index   283

Preface

Why fractals, natural surfaces and scattering? These are actually two questions compressed in a single statement and should be rephrased as follows: why fractals and natural surfaces? And why (electromagnetic) scattering by such surfaces?

Let us consider the first question, which is essentially related to the appropriate model of the natural surfaces, i.e., surfaces that landscape our planet (and the other ones, too) and are not man-made. There is increasing experimental evidence that for such a model the Euclidean geometry is not the appropriate one: a better match is obtained by employing the fractal geometry. The reason is that forces that model natural surfaces: gravity and microgravity, tensions, frictions, vibrations, erosions, thermal and freezing gradients, chemical reaction, etc; and periodic and a-periodic happenings: seasons and vegetation changes, sun, wind, rain, snow, slides, subsidence, etc. generate surfaces whose topological dimension is larger than 2; i.e., larger than the Euclidean one. Loosely speaking, the corrugations and microondulations impressed by the natural forces tend to expand the surface into the surrounding volume, thus "extending" its topological dimension in the range from 2 (Euclidean surface) to 3 (Euclidean volume). Such a geometry is well described by the fractal geometry, which considers surfaces whose topological dimension is equal to $D = 3 - H, 0 < H < 1$ being the Hurst exponent. The conclusion is that natural surfaces need to be modeled by an "ad hoc" geometry, the fractal one.

Assume now that the fractal geometry has been adopted, and the electromagnetic theory is used to compute the field scattered by the modeled surface illuminated by prescribed sources. Then, the second question comes up: why invest time and effort for setting up computational algorithms and procedures, while these are already available in the existing literature? In other words, these techniques have already been developed and applied to the conventional (Euclidean) models of the natural surfaces: it seems that one can just change the model and make use of the same techniques. The appropriate answer to this question is a bit elaborate and is presented hereafter.

Fractal geometry is mathematical abstraction of fractal physics: it exhibits properties (for instance, self-affinity) on all scales and does not allow the derivative operation at any point; surface fractal corrugations possess power spectra that diverge in the low-frequency regime (infrared catastrophe) and exhibit non-stationary correlations functions. Use of the mathematical fractals to model natural surfaces would make any scattering computation totally intractable. But natural surfaces are observed, sensed, measured, and represented via instruments that are, for their intrinsic nature, bandlimited. Accordingly, the mathematical fractals may, or must be bandlimited, thus generating the physical fractals that recover most, if not all, the properties needed to manage them in the electromagnetic scattering theory. In spite of this, it is not immediate to transfer known conventional scattering computational techniques to the new geometry: a complete rephrasing is necessary.

This book addresses in detail all previous questions and provides all needed information to tackle the problem of electromagnetic scattering by fractally modeled natural surfaces. It is divided in two segments of four chapter each, the first segment being propedeutic to the second one.

The first four chapters present the fractal geometry and address in detail the bandlimitation issue, thus examining the properties of physical fractals. Comparison of the fractal model with the conventional one is discussed as well, and the fundamentals of the existing methods for (electromagnetic) scattering computation is included.

The last four chapters are based on the first four ones and present the full theory of scattering from fractal surfaces, introduce the pertinent parameters, mainly the scattering cross section, and elucidate the physical implications of obtained results. The result is a complete monography on the subject.

The authors are grateful to all the scientific community with whom they interacted over the years and that "modeled" their ideas, developments and conclusions on this fascinating research area.

Giorgio Franceschetti
Daniele Riccio

CHAPTER 1

The Scattering Problem

1.1. Introduction and Chapter Outline

This chapter presents the rationale and the motivations for this book on scattering from natural surfaces. Distinction is made between the classical and the fractal description of the surfaces; its relevance to the evaluation of the scattered field is elucidated. Regular and predictable stochastic processes are introduced; their use for the evaluation of the scattered field within different electromagnetic methods is presented in detail.

Definition of the scattering problem is provided in Section 1.2, whereas motivations to obtain a solution for it are given in Section 1.3. Geometric models for the scattering surface are introduced in Section 1.4, where it is also discussed how they enter into the evaluation of the scattered field.

The scattering problem is then categorized with respect to both the model employed for the scattering surface and the electromagnetic method applied to find the scattered field: deterministic and stochastic models for the scattering surface are introduced and compared in Section 1.5; deterministic and stochastic evaluations for the scattered field are presented and compared in Section 1.6. Then, depending on the selected surface model and electromagnetic method, analytic and numerical solution can be obtained and are introduced in Section 1.7.

Section 1.8 deals with the analytic evaluation in closed form of the scattered field, which is of main concern in this book according to the motivations presented in Section 1.3.

The book outline is introduced in Section 1.9.

Key references and suggestions for further readings are reported in Section 1.10.

1.2. The Scattering-Problem Definition

In the electromagnetic-field theory, *scattering problems* arise whenever interaction of electromagnetic fields with matter takes place. Among these problems, scattering from *natural surfaces* is of paramount relevance, because it plays a fundamental role in wave propagation and remote sensing.

Consider a natural surface separating two homogeneous media of semi-infinite extent (Figure 1.1). The key point is that the surface is *natural*—that is, *not* man-made. Accordingly, it lacks the clear-cut geometric properties and features that are typical of man-made structures. And it is difficult, if not impossible, not only to find a general closed-form expression for the scattered field, but even to generate a reasonably accurate model for the surface itself. Attention is then devoted to find a convenient approach to this issue.

The problem to be handled can be defined as follows. When the natural surface is illuminated by an electromagnetic field, a scattered field is generated. Accordingly, the total field is decomposed into the sum of two ones, as detailed in the following paragraphs.

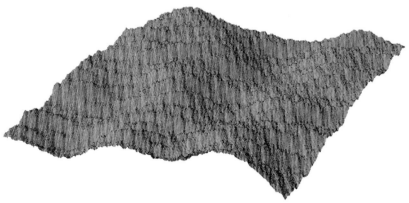

FIGURE 1.1 Drawing of a portion of a natural surface separating two homogeneous media of semi-infinite extent.

The *incident field* is the one that would be present if the medium were homogeneous, and coincident with the free-space one. Accordingly, the scattering surface is no longer present. This incident field is usually taken as a single, possibly plane, wave: field expansion in convenient bases provides both rationale and techniques to handle the problem in the most general case.

The *scattered field* is a solution of the source-free Maxwell equations, satisfying radiation conditions at infinity. It is usually represented by different expressions inside and outside the natural surface.

The incident and scattered fields together provide the actual electromagnetic field; continuity conditions of the tangential component of the total fields on the natural surface must be enforced.

Evaluation of the scattered field is obtained by following analytic as well as numerical techniques. In the case of natural surfaces, this is accomplished by a two-step procedure: first the surface is modeled, then a method for evaluating the scattered field is developed. This second step is strongly related to the outcome of the first step. The two steps are somehow coupled: the first step mainly implies to make choices with subsequent mathematical elaborations; the second is mainly a matter of electromagnetic theory.

1.3. Motivations

Many important applications are connected to the solution of the presented scattering problem: electromagnetic-wave propagation, radio communication, remote sensing, radar detection, electromagnetic diagnostics, electromagnetic imaging. All these disciplines widely benefit from availability of affordable scattering methods able to efficiently evaluate the interaction of the electromagnetic field with natural surfaces.

A simple closed-form formulation of the scattered field is a necessary prerequisite to understand the scattering rationale, to identify key electromagnetic and surface parameters relevant to the scattering mechanism, to plan and design scattering sensors and radio-communications instruments, to predict scattering-phenomena effects, to provide the rationale to simulate the electromagnetic fields scattered from rough surfaces, and to develop an affordable tool to interpolate and extrapolate scattered fields from measurement data.

Scattering methods provide a solution to the *direct problem*, aimed at evaluating the electromagnetic field whenever the surface is fully described.

But the *inverse problem* is of interest, too. It consists of estimating surface properties from scattering data: direct procedures are the background, and provide unique guidelines to develop scattering inverse methods. Again, clarity and simplicity of direct methods are the necessary prerequisite to developing affordable inverse methods.

1.4. Surface Models and Electromagnetic Methods

In this book, the first step toward evaluating the scattered field is referred to as *surface-modeling*; the second one is referred to as *electromagnetic-evaluation procedure*. This is different from what is common in the relevant literature, where these two steps are usually unified and referred to as the electromagnetic model of the problem. We make reference to this approach as the *classical one*: this term is somewhat motivated by the almost standard description used in modeling natural surfaces, so that the major emphasis is on the scattered-field computation.

In this book, the above-quoted two steps are considered separately, because alternative surface models are introduced. Geometric modeling of natural surfaces is provided by means of the *fractal geometry*, as widely suggested by geologic and oceanographic studies: in other words, the proposed geometric models are taken from advanced studies in the field of natural sciences and mathematics. It is noted that fractal models of natural surfaces do not hold those mathematical properties required to apply classical electromagnetic methods in order to analytically evaluate the scattered field in closed form. Hence, the scattered-field evaluation procedures are here completely restated according to the characteristics of fractal surfaces, as detailed in Chapters 5 through 8.

Fractal geometry is illustrated in several excellent books in mathematical sciences. Some of them deal with natural-surfaces modeling. Analytic models reported here are able to handle a variety of natural surfaces: bare and moderately vegetated soils, as well as ocean surfaces. Forested areas are excluded from the presented analysis, because only surface, and not volume scattering, is considered here.

Electromagnetic-scattering methods based on fractal models for the scattering surfaces can be found only in journal papers and a few book chapters. Accordingly, this book fills a void in the literature, by providing a common mathematical and electromagnetic approach to unify and enlarge the applicability of scattering theories to fractal surfaces.

1.5. Deterministic versus Stochastic Models for the Natural Surfaces

The geometric shape of natural surfaces can be represented by means of a deterministic function as well as a stochastic process.

If a specific or assigned natural surface is under study, and if scattering from this particular surface must be evaluated, then a deterministic function may be used to model the surface geometry. Conversely, if a class of natural surfaces is considered and some average or significant parameters of the expected scattered field from that class of natural surfaces is desired, then a stochastic process may be appropriately used to model the considered whole class of natural surfaces.

The choice between these two very different descriptions relies on their expected use. In this section, the rationale to use deterministic or stochastic description for the natural surfaces is discussed. Final comments are devoted to comparing classical- and fractal-based approaches.

1.5.1. Surface Deterministic Models

In the deterministic description of a rough surface, the geometric shape may be modeled by means of real, possibly single-valued, functions of two independent space variables. These functions may hold some relevant properties such as continuity and derivability, and usually belong to some class of real functions.

From a general viewpoint, scattering from canonical structures—including simple shapes, periodic structures, and made-made objects—relies on a deterministic description of the surface of these bodies. For simple geometric shapes, this evaluation is analytically performed in closed form, in terms of the geometric and the electromagnetic parameters of the surface.

A deterministic description is seldom used in the classical analytic evaluation of electromagnetic scattering from natural surfaces, because its use does not generally lead to closed-form expressions for the solution. A noticeable exception is provided by surfaces modeled by almost-periodic functions, whose scattering properties, although cumbersome, are amenable to an analytic solution.

A reasonable deterministic fractal description for natural surfaces is provided by the *Weierstrass-Mandelbrot* (WM) function. Moreover, randomization of the finite number of parameters of the WM function provides an easy generation model for a stochastic surface.

Finally, deterministic description is also used to represent natural surfaces whenever a numerical evaluation of the scattered field is in order.

1.5.2. Surface Stochastic Models

In the stochastic description of rough surfaces, the geometric shape may be modeled by means of a *stochastic process* of two independent space variables. In this case, a Cartesian coordinate system is employed, $z(\mathbf{r}) = z(x, y)$, $\mathbf{r} \equiv x\hat{\mathbf{x}} + y\hat{\mathbf{y}}$, where $z = 0$ is the mean plane and $z(x, y)$ describes its stochastic corrugations. It is worth recalling that a stochastic process $z(\mathbf{r})$ is a rule for assigning to every outcome ς of the *statistical ensemble* a function $z(\mathbf{r})$. Thus, a stochastic process is an ensemble of functions of the space variables depending on the parameter ς.

Alternatively, a stochastic process can be read as a function $z(x, y, \varsigma)$ of three variables: the domain of (x, y) is R^2, the domain of ς is the set of all the experimental outcomes. If the position, $\mathbf{r}$, over the surface is variable, and the realization, ς, is fixed, then z is a *sample space function*; if the position, $\mathbf{r}$, over the surface is fixed, and the realization, ς, is variable, then z is a *random variable*; if both position, $\mathbf{r}$, and realization, ς, are fixed, then z is a number.

In this book, stochastic processes are simply indicated as $z(\mathbf{r})$ whenever used to describe natural-surface shapes.

Obviously, the field scattered from a stochastic surface exhibits itself a stochastic behavior. For each space position, the scattered field is a random variable in the phasor domain, and a stochastic process of the independent time variable in the time domain. This point is detailed in the Section 1.6.2.

An important classification is between *regular* and *predictable* stochastic processes. The difference is that a regular stochastic process consists of an ensemble of functions that cannot be described in terms of a finite set of parameters, in contrast to a predictable stochastic process consisting of an ensemble of functions that can be completely specified by a finite set of parameters. The ensembles of regular stochastic process are detailed via appropriate stochastic distributions depending on an appropriate finite set of stochastic parameters; this does not happen for predictable processes.

As an example, consider the height profile $z(x, y)$ of a rough surface. If the probability distributions of the random process are prescribed, then

1.5. Deterministic versus Stochastic Models for the Natural Surfaces

a regular stochastic process is in order. Conversely, the process is predictable if $z(x, y)$ is expanded in a series of functions that contain statistical parameters, specified in terms of appropriate probability distributions.

Some interesting properties of regular and predictable processes are in order. These properties are fundamentals for selecting the type of the stochastic model to be conveniently used.

For predictable stochastic processes, knowledge of a sample function $z(\mathbf{r})$ on a subdomain may lead to predicting the whole sample function. This is not true for a regular stochastic process. Conversely, for regular stochastic processes, knowledge of a sample function $z(\mathbf{r})$ may lead to reconstructing the statistics of the entire process. This is not true for a predictable stochastic process.

In the following paragraphs, both regular and predictable stochastic processes are used for the natural-surface shape modeling.

Classical rough surfaces are generally introduced as regular stochastic processes. Use of appropriate synthesis procedures may generate a realization of the stochastic process, leading to a deterministic classical surface. As an alternative, the realization may be experimentally obtained by means of in situ measurements.

Fractal stochastic surfaces are introduced by using either regular or predictable stochastic processes. The *fractional Brownian motion* (fBm) model is a popular fractal regular stochastic process. Synthesis procedures may then be used to obtain an element of the ensemble—that is, a deterministic surface. The WM function is a popular fractal deterministic function. Its coefficients can be easily randomized to generate random surfaces. This amounts to constructing a predictable stochastic fractal process.

For a visual comparison, examples of fractal and classical surfaces are provided in Figure 1.2. A fractal surface is presented in the first column, along with its contour lines (corresponding to five horizontal cuts) and a generic vertical cut: the classical counterparts are reported in the second column. The lighter the gray level of the area between successive contour lines, the higher the surface. Beyond any mathematical or physical explanation, it is evident that the visual comparison between the surfaces models presented in Figure 1.2 and the natural surfaces is definitively in favor of the fractal models: the fractal surface exhibits details on any observation scale, whereas the classical one appears too smooth at the smallest scales; this is confirmed by the surfaces drawing and the surfaces' horizontal and vertical cuts.

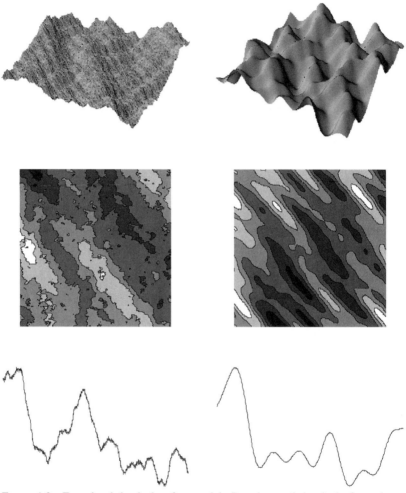

FIGURE 1.2 Fractal and classical surface models. Drawings and plots in the first column of the figure are referred to the fractal model, whereas their counterparts for a classical model are reported in the second column. In the first row of the figure the fractal and classical surfaces are drawn. In the second row the corresponding contour plots (on five levels) are presented. In the third row an example of surfaces' vertical profiles is shown: the trace of the vertical profiles is a horizontal line passing to the center of the contour plots.

1.6. Deterministic versus Stochastic Evaluation for the Scattered Field

The electromagnetic field scattered from a natural surface can be represented either as a deterministic function or as a stochastic process; in the former case, the field is represented by its value; in the latter, by a random variable.

The choice between these two very different descriptions relies on their expected use, on the possibility to get appropriate—analytical or numerical—solutions to the scattering problem, and is related strictly to the corresponding choice performed to model the scattering surface. In this section, the rationale to consider deterministic or stochastic descriptions for the scattered field is discussed.

1.6.1. Scattered-Field Deterministic Descriptions

The electromagnetic field scattered from a natural surface can be evaluated in a deterministic form if the surface description is deterministically provided: in such a case, the problem consists of computing the field scattered from the prescribed specific surface.

As discussed in the previous section, in the analytic evaluation of the scattered electromagnetic field, a deterministic description for classical surfaces is not used. This is because any classical-surface model, somehow approaching the shape of the natural surface, is too much involved and does not lead to an analytic closed-form solution. In some cases, the only partial exception to this rule seems to be provided by almost-periodic surfaces.

Conversely, a reasonable deterministic fractal description for natural surfaces is provided by the WM function. In this book, it is shown that a deterministic description for the electromagnetic scattered field from WM surfaces can be analytically evaluated in closed form (see Chapters 5 and 7). Obviously, randomization of the finite number of parameters of the WM function easily allows the evaluation of the scattered field as a predictable stochastic process. It also turns out that the ensemble element of the scattered field can be obtained when a predictable stochastic process model for the surface is adopted. Conversely, only statistical parameters of the scattered field can be evaluated if a regular stochastic process model is used.

Deterministic descriptions for the scattered field are of paramount relevance if numerical methods are in order. Any numerical evaluation of the scattered field leads to a deterministic description; as a matter of fact, numerical methods deterministically evaluate the scattered field for a prescribed specific surface. When stochastic-process modeling for the scattering surface is used, as a first step the elements of the scattering surface ensemble are generated; then the scattered field for each ensemble element is numerically evaluated. Multiple application of the procedure can be later used to determine the stochastic behavior of the scattered field by implementing appropriate averages on obtained numerical solutions.

1.6.2. Scattered-Field Stochastic Descriptions

The field scattered from a natural surface can be analytically evaluated in a stochastic form if the surface description is provided in stochastic terms: in such a case, the problem consists of determining the stochastic characterization of the field scattered from a prescribed class of surfaces. To this end, the distinction between regular and predictable processes, Section 1.5, plays an important role. In the case of predictable processes, a particular choice of the value of the finite number of its describing parameters fully determines the ensemble element. Accordingly, this element is analytically and deterministically individuated, and computational techniques can be applied, in principle, for the evaluation of the scattered field. This is an element of the ensemble of the scattered field associated to the considered surface, and its statistical properties may be further explored by operating on the ensemble set. Conversely, when a regular process is in order, each ensemble element is by itself described only in a statistical way. Accordingly, no analytic description of the element is available, and only statistical properties of the scattered field may be directly explored.

For both cases of the analytic and numerical evaluation of the scattered field, the final result is either a phasor or a time-dependent analytical or numerical expression. In the phasor domain, the vector stochastic process $\mathbf{E}(\mathbf{r})$ is a rule for assigning to each ensemble element ς of the surface a function $\mathbf{E}(\mathbf{r})$ of three independent space variables; at any space point, the scattered field is a vector random variable. In time domain, a further temporal independent variable t is added, so that the vector stochastic process $\mathbf{e}(\mathbf{r}, t)$ depends on four independent variables; at any space point, the scattered field is a stochastic process with one independent variable—that is, time. In the following paragraphs, the analysis is driven in the phasor domain; extension to the temporal domain may be gained by application of standard Fourier Transform (FT) techniques.

Thus, a stochastic electromagnetic field is an ensemble of functions of the space variables depending on the parameter ς. If the space position, $\mathbf{r}$,

1.6. Deterministic versus Stochastic Evaluation for the Scattered Field

is variable, and the realization, ς, is fixed, then **E** is a vector function of the variable **r**; if the space position, **r**, is fixed, and the realization, ς, is variable, then **E** is a vector random variable; if both the position, **r**, and the realization, ς, are fixed, then **E** is a vector.

There is obviously a relevant link between the stochastic description of the scattering surface and the electromagnetic field: the scattered field evaluated in closed form inherits the class of the stochastic process from the surface. In particular, if the surface is described by means of a regular stochastic process, then the scattered field is a regular stochastic process; similarly, if the surface is described by means of a predictable stochastic process, then the scattered field is a predictable stochastic process. Representation of the scattered field for predictable and regular processes deserves some comments.

Predictable stochastic-process modeling can be used to describe fractal surfaces: this happens, for instance, whenever the WM function with random coefficients is employed. In this case, the scattered field, and not only its stochastic parameters, can be evaluated in closed form (see Chapters 5 and 7). If relevant averages of the scattered field are of interest, they can be evaluated (in some cases in closed form) by employing the obtained closed-form solution for the scattered field.

Regular stochastic processes are used to describe classical surfaces as well as fBm fractal-model surfaces: in both cases, the scattered field is a regular process, and a closed-form solution for the scattered field cannot be gained—but this does not imply that scattering computational procedures are not implemented or necessary. As a matter of fact, an analytic formulation of the scattered field must be developed; the stochastic parameters of the field are then analytically evaluated in closed form, by operating appropriate averages on the scattered-field analytical formulation. Those averages obviously require knowledge of the prescribed statistical parameters of the height profile, and this step imprints the statistics of the surface on the scattered field. In conclusion, in the case that regular processes are employed, the field can be described only by means of its stochastic parameters. Usually only a few of them are evaluated. The mean field is the first: however, for natural surfaces whose roughness is not marginal, the mean scattered field is usually zero. Then the field variance, related to the scattered power, is evaluated. For remote-sensing application, the scattered power is employed to evaluate the radar cross section, which can be properly normalized to the geometric area, thus leading to the scattering coefficient. Evaluation of these two relevant averages, scattered mean field and power, is often sufficient to characterize the scattered field at a single point in space.

It is important to note that a full description for the scattered field requires evaluating much more than some relevant averages: as a matter of fact, knowledge of single-point and multipoint (at any order) joint probability-density function (pdf) of the stochastic field are required instead. To address this item, some comments are listed in the following sections.

In most applications, the first- and second-order statistics are usually sufficient for a satisfactory description of the single-point scattered-field behavior. This happens, for instance, whenever the central-limit theorem applies and the modulus of the scattered field turns out to be Rayleigh distributed. In this case, the scattered field in any single space point is a stochastic variable fully described at any order by one parameter, this latter parameter being related to the scattered power. But this is not always the case: for instance, some marine surfaces exhibit a scattered field whose modulus is not Rayleigh distributed; in such a case, the full stochastic characterization (in principle at any order) of the scattered field is required to provide useful information on the scattering surface. It is important to note that in this case, the required higher-order statistics of the single-point scattered field are difficult to find in the current literature. This is mainly because these averages are difficult to obtain analytically in closed form.

In some cases, multipoint stochastic characterization of the scattered field is of interest. In particular, the two-point stochastic characterization describes the scattered field space and/or time correlation. In the former case, a convenient meaningful statistical parameter for the scattered field is its correlation length; in the latter, its correlation time. In general, the covariance matrix of the field should be computed; this is a more difficult task then finding the single-point statistics. In remote-sensing applications, its importance is confined to interferometric applications; in cellular propagation, to wideband signals. This book is confined mostly to the single-point scattered-field characterization and applications.

1.7. Analytic versus Numerical Evaluation of the Scattered Field

An exact closed-form solution for the electromagnetic field scattered from natural surfaces does not exist: any approach relies on some approximations.

1.7. Analytic versus Numerical Evaluation of the Scattered Field

The corresponding solution is approximate as well. In spite of these limitations, classical analytic methods often shed light on the scattering mechanism and sometimes provide the rationale for interpreting the scattered data; however, they are rarely satisfactory with respect to the applications reported in Section 1.3. Then new scattering methods are continuously proposed, with the aim of improving their applicability; but usually small improvements in classical approaches are paid for by a large increase in methods complexity.

In this book, analytic methods to evaluate the scattered field in closed form, tailored to the fractal models, are presented. These methods are accommodated to the fractal models, discussing the involved approximations: each method allows expressing the scattered field as a function of the surface fractal and electromagnetic parameters, as well as the sensor geometric and electromagnetic parameters. A complete list of these parameters is now presented in detail: this is crucial to identifying the potentiality inherent to any scattering model.

Surface parameters are in order. In Chapter 3, it is shown that fractal models involve two fractal parameters; for instance, fractal dimension and topothesy, or alternatively, Hurst exponent and surface-increments standard deviation can be used. Finally, if spectral representations are considered, spectral amplitude and slope are used. The electromagnetic parameters, permittivity and conductivity of the homogeneous medium below the surface are grouped in the complex dielectric constant in the usual nonmagnetic-material case. Coming to the sensor parameters, the narrowband bistatic radar is considered: the usual parameters are the geometric coordinates of the transmitter and the receiver as well as the radar carrier frequency.

A comment on classical approaches is in order. In Chapter 2, it is shown that classical models for the scattering surface involve at least two parameters: usually the surface standard deviation and the correlation length. However, the two parameters lead to models that very poorly account for natural-surfaces properties, and their limitation can be mitigated by increasing the parameters' number. This (marginal) improvement in the model is paid for by a significant complication in the evaluation of the scattered field.

Classical procedures to evaluate the scattered electromagnetic field are available in several excellent books. In some cases, these analytic classical procedures are also briefly referred to, and this is for two reasons. First, to underline why they cannot be applied to the fractal surfaces; and second, to provide a comparison with the results based on fractal geometry.

Numerical solutions to the problem are also available. In this case, the scattered field is not obtained in a closed form, and most of the motivation discussed in Section 1.3 cannot be fulfilled. However, numerical procedures can be used to test the analytic ones. It is important to underline that also the numerical procedures lead to approximate solutions to the scattering problem. Hence, the numerical approach also needs some theoretical discussions to provide its assessment with respect to the actual data.

1.8. Closed-Form Evaluation of the Electromagnetic Field Scattered from a Natural Surface

Scattering studies related to classical natural surfaces, modeled as rough geometric objects, date back to the early 1950s. Different methods for the scattered-field evaluation have been developed, each based on different approximations, hence with a different range of validity. Among them, the most popular ones are the *Kirchhoff Approximation* (KA); the *Extended-Boundary-Condition Method* (EBCM), with the derived *Small-Perturbation Method* (SPM); and, more recently, the *Integral-Equation Method* (IEM). These techniques are by now well established, and are widely available in the current literature. They can be quite easily combined with classical statistical models, which describe natural rough surfaces by means of stationary stochastic two-dimensional processes, with given *probability-density function* (pdf) (usually Gaussian) and *correlation function* (usually Gaussian, exponential, or a combination of the two).

Evaluated scattered fields are in excellent agreement with numerical simulations and laboratory measurements (performed by using artificial rough surfaces consistent with above surface models). However, their comparison with measured data over actual natural surfaces is often less encouraging, and this is probably due to the inadequacy of the adopted surface models. As a matter of fact, natural surfaces exhibit appropriate statistical scale-invariance properties that are not met at all by surface classical models. In Figure 1.3, the concept of this appropriate scale invariance, termed self-affinity, is elucidated: the natural-surface profile at the top of the figure is zoomed by a factor of 10 in the second row, and by a factor of 100 in the third row. Actually, any zooming in reveals finer details, and the surface appears rougher at the smallest scales. A little thought on this point leads to the conclusion that this is in agreement with experience: a rock

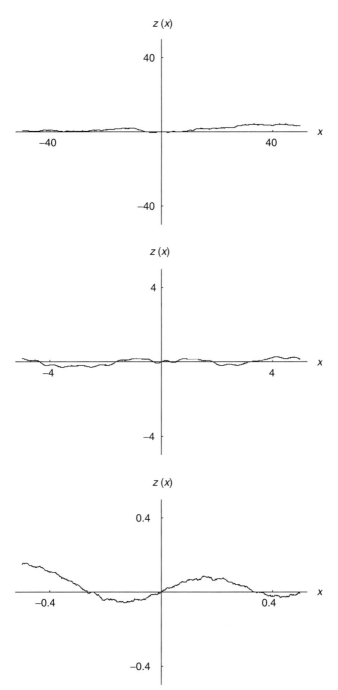

FIGURE 1.3 Self-affinity properties of natural and fractal profiles. The surface profile at the top of the figure is zoomed by a factor 10 in the second row and by a factor 100 in the third row. Any zooming in reveals finer details of the surface which appear rougher at smallest scales.

usually appears relatively, not absolutely, rougher than a mountain. Then again, a valid candidate for the natural surface cannot be found among the classical models, or equivalently, a better natural-surface description can be obtained by using the fractal models.

Fractal geometry was applied in the mid 1970s in order to provide a mathematical tool to deal with the complex and irregular shape of natural objects. The effectiveness of fractals to describe natural surfaces has been demonstrated in a very impressive way by surprisingly realistic computer-generated synthetic landscapes (see Figure 1.4). One of the reasons for this success is the ability of fractal models to properly account for the statistical *scale-invariance* properties (in particular, self-affinity) of natural surfaces whose graphical evidence is shown in Figure 1.3 (for mathematical details,

FIGURE 1.4a   A computer generated natural landscape.

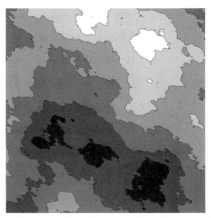

FIGURE 1.4b   Contour plot of the natural landscape in Figure 1.4a.

1.8. Closed-Form Evaluation of the Electromagnetic Field

FIGURE 1.4c   Computer generated natural landscape typical of a seaside area.

FIGURE 1.4d   Computer generated natural landscape typical of a mountainous area.

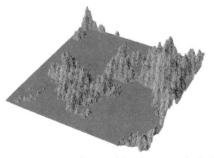

FIGURE 1.4e   Computer generated natural landscape typical of an archipelago.

see Section 3.2.3). Accordingly, fractal models, and in particular the fBm one, have been used recently in different disciplines to describe natural surfaces. Although use of fractal models in electromagnetic scattering is not straightforward, some successful attempts have been reported recently. Such approaches can be roughly divided into two categories: the first one includes methods in which the fBm model is approximated by using the WM band-limited function; the second one includes methods that directly use the definition of the fBm process.

The main advantage of using the WM function is that an analytic expression of the scattered field can be obtained. However, this expression is quite involved, and it is not possible to simply analytically evaluate the (expected) scattered-power density. On the other hand, direct use of the fBm definition allows obtaining closed-form expressions of the scattered-power density, at least by using either the KA or the SPM approach. The solutions are not too involved, and allow an easy evaluation of the scattered-power density dependence on the surface fractal parameters. The disadvantage is that only the second-order statistics of the scattered field can be evaluated, and not the scattered field itself. In spite of their simplicity, all these fractal-based solutions to the scattering problem always lead to results that are in good agreement with measured data.

Use of fractal models in connection with IEM is a delicate point. Currently, a closed-form solution for the scattered-field density or scattered-power density does not exist, because some mathematical requirements involved in the IEM approach are not allowed whenever a fractal surface is considered.

1.9. Book Outline

Chapter 1 provides the motivations and the rationale for a book on scattering from natural surfaces modeled by means of fractal geometry.

Chapter 2 presents classical surfaces as they are employed in classical-scattering models.

Chapter 3 presents fractal surfaces as they are used in scattering methods appropriate to these surfaces. Both fBm and WM models are analyzed, and the link between them is highlighted. Fractal parameters used to characterize natural surfaces are defined and illustrated, and their relations with classical-surface parameters are presented.

Chapter 4 sets up evaluation procedures for computing the scattered electromagnetic field. The general theoretical background is presented before any choice on the evaluation techniques and the surface model is detailed.

Closed-form solutions for the field scattered from fractal surfaces are presented in Chapters 5 through 8. In each chapter, a different electromagnetic method or a different geometrical fractal model is adopted. In Chapters 5 and 6, the analytic solution is found within the KA; conversely, the EBCM is followed in Chapters 7 and 8. In Chapters 5 and 7, the geometric-surface model falls within the class of random predictable fractal processes; conversely, in Chapters 6 and 8, the surface geometric model make reference to random regular fractal processes.

1.10. References and Further Readings

Suggested references consist of books and journal papers, which are classified into two different lists at the end of the book. Only most of the books are referred to in the reference and further reading sections of each chapter.

Many books have been published that present theory and results relevant to scattering from classical random surfaces: only the most popular ones are reported. References include books on random surfaces (Beckmann and Spizzichino 1987), microwave-scattering theory devoted to remote sensing (Fung 1994; Tsang, Kong, and Shin 1985; Ulaby, Moore, and Fung 1982), scattering and wave propagation (Ishimaru 1993), and nonhomogeneous media (Chew 1995).

Contributions on scattering from fractal shapes are available only in some journal papers (see "References Involving the Authors of This Book" in Appendix C) and some chapters in books (Jaggard 1990).

Fundamentals on stochastic processes are available (Papoulis 1965), as well as on fractal geometry (Falconer 1990; Feder 1988; Mandelbrot 1983).

Other references deal with mathematical issues (Abramowitz and Stegun 1970; Gradshteyn and Ryzhik 1980; Prudnikov, Brychkov, and Marichev 1990).

CHAPTER 2

Surface Classical Models

2.1. Introduction and Chapter Outline

This chapter presents the classical geometric models used to describe natural surfaces, with a special emphasis on those that are appropriate for use in the area of electromagnetic scattering theory.

Deterministic as well as random representations are introduced, along with relevant discussions on their use. If a stochastic surface model is employed for the surface, then the scattered field is also stochastic; hence, the choice of the surface representation is related to the desired, or allowed, evaluation of the scattered field.

Fundamentals on stochastic processes are preliminarily introduced in Sections 2.2 through 2.4. Key definitions, relevant averages, and a short discussion on stationary processes are presented in Section 2.2, wherein reference is made to the space domain. Section 2.3 is devoted to introducing the stochastic-process spectral characterization whenever stationary processes are in order. The descriptions in spectral and space domain are then linked. Key relationships introduced in Sections 2.2 and 2.3 are specialized in Section 2.4 to the relevant case of isotropic surfaces.

Classical models for rough surfaces are then presented in Sections 2.5 through 2.6 and Appendix 2.A. Section 2.5 is devoted to presenting the first-order characterization, and focuses on the popular Gaussian case. Section 2.6 introduces for rough surfaces the second-order characterization; autocorrelation functions that allow closed-form evaluation of their spectral counterpart are presented in detail in Appendix 2.A. The physical counterpart of the parameters entering into the first- and second-order surface characterizations is reported in Section 2.7.

Finally, the rationale for selecting the classical rough-surfaces models to evaluate the scattered electromagnetic field is reported in Section 2.8: emphasis is placed on the required order description for the surface shape.

Key references and suggestions for further readings are reported in Section 2.9.

2.2. Fundamentals of Stochastic Processes

Fundamentals of stochastic processes are reported in this section. Wider discussions can be found in many books and in the specific literature of this branch of mathematics; the few notes reported here have the intent only to introduce the reader to notations and symbols used in this book.

Characterization of stochastic processes is briefly introduced. Cumulative-distribution functions and probability-density function (pdf) are presented. Some relevant parameters commonly used to synthetically describe stochastic processes are defined. Several scalar processes are of interest in this book: for instance, the surface height and any component of the electromagnetic phasor or spectral field. To describe the process, reference can be made to a complex scalar process, $z(\mathbf{r})$, defined in C^n, $n = 1, 2, 3$. The complex conjugate product is implemented, because $z(\mathbf{r})$ can attain complex values; in the simplest case of the height profile, $z(\mathbf{r})$ is real, $\mathbf{r} = x\hat{\mathbf{x}} + y\hat{\mathbf{y}}$—that is, $n = 2$, and the complex conjugate may be omitted.

2.2.1. Stochastic Processes: Definition

A stochastic process $z(\mathbf{r})$ is an uncountable infinity of random variables, one for each $\mathbf{r}$. In the modeling of surface geometric properties, which is of concern in this book, $\mathbf{r}$ is the vector coordinate of the plane, and $z(\mathbf{r})$ is the random height perturbation of the surface.

A stochastic process is determined to the first order if, for each ζ and for each $\mathbf{r}$, the *first-order Cumulative-Distribution Function* (CDF), $F(\zeta, \mathbf{r})$, is assigned:

$$\Pr\{z(\mathbf{r}) \leq \zeta\} \stackrel{\Delta}{=} F(\zeta, \mathbf{r}), \quad (2.1)$$

where $\Pr\{\cdot\}$ means probability, and ζ is an independent random variable.

2.2. Fundamentals of Stochastic Processes

The derivative of $F(\zeta, \mathbf{r})$ with respect to ζ provides $p(\zeta, \mathbf{r})$, the *first-order probability-density function* (pdf),

$$p(\zeta, \mathbf{r}) \triangleq \frac{\partial F(\zeta, \mathbf{r})}{\partial \zeta}, \qquad (2.2)$$

which can be used, as an alternative to $F(\zeta, \mathbf{r})$, to determine the stochastic process.

A stochastic process is determined to the second order if the *joint-second-order CDF*, $F(\zeta_1, \zeta_2; \mathbf{r}_1, \mathbf{r}_2)$, is assigned:

$$\Pr\{z(\mathbf{r}_1) \leq \zeta_1, z(\mathbf{r}_2) \leq \zeta_2\} \triangleq F(\zeta_1, \zeta_2; \mathbf{r}_1, \mathbf{r}_2) \qquad (2.3)$$

whereas the corresponding *joint-second-order pdf* is

$$p(\zeta_1, \zeta_2; \mathbf{r}_1, \mathbf{r}_2) \triangleq \frac{\partial^2 F(\zeta_1, \zeta_2; \mathbf{r}_1, \mathbf{r}_2)}{\partial \zeta_1 \partial \zeta_2}. \qquad (2.4)$$

Higher-order characterization of the stochastic process can be introduced by following the rationale illustrated in Equations (2.1) through (2.4).

A stochastic process is fully determined if, for each n, the *joint n-th order CDF*, $F(\zeta_1, \ldots, \zeta_n; \mathbf{r}_1, \ldots, \mathbf{r}_n)$, is assigned:

$$\Pr\{z(\mathbf{r}_1) \leq \zeta_1, \ldots, z(\mathbf{r}_n) \leq \zeta_n\} \triangleq F(\zeta_1, \ldots, \zeta_n; \mathbf{r}_1, \ldots, \mathbf{r}_n), \qquad (2.5)$$

the corresponding *n-th order pdf* being

$$p(\zeta_1, \ldots, \zeta_n; \mathbf{r}_1, \ldots, \mathbf{r}_n) \triangleq \frac{\partial^n F(\zeta_1, \ldots, \zeta_n; \mathbf{r}_1, \ldots, \mathbf{r}_n)}{\partial \zeta_1 \ldots \partial \zeta_n}. \qquad (2.6)$$

2.2.2. Stochastic Processes: Relevant Averages

In many applications, the complete knowledge of a stochastic process is not fully required, and it can be assigned up to a prescribed order. This simplified approach is efficiently applied whenever only some statistical averages relevant to the stochastic process are of interest: these quantities allow the evaluation of some relevant meaningful physical observables such as the mean scattered field and scattered-power density. Usually, for a random rough surface, only the first-order pdf and some averages related to the joint-second-order pdf are used.

Most relevant averages for a stochastic process are presented hereafter.

The *statistical mean* of the stochastic process is defined by means of the first-order pdf,

$$\mu(\mathbf{r}) \stackrel{\Delta}{=} \langle z(\mathbf{r}) \rangle \stackrel{\Delta}{=} \int_{-\infty}^{\infty} \zeta p(\zeta, \mathbf{r}) \, d\zeta. \tag{2.7}$$

The statistical mean can be recognized as the first-order moment of the stochastic process, the generic m-order one being defined as

$$\mu_m(\mathbf{r}) \stackrel{\Delta}{=} \langle z^m(\mathbf{r}) \rangle = \int_{-\infty}^{\infty} \zeta^m p(\zeta, \mathbf{r}) \, d\zeta. \tag{2.8}$$

The *autocorrelation* of the stochastic process is defined by means of the second-order pdf:

$$R(\mathbf{r}_1, \mathbf{r}_2) \stackrel{\Delta}{=} \langle z(\mathbf{r}_1) z^*(\mathbf{r}_2) \rangle = \int_{-\infty}^{\infty} \int_{-\infty}^{\infty} \zeta_1 \zeta_2^* p(\zeta_1, \zeta_2^*; \mathbf{r}_1, \mathbf{r}_2) \, d\zeta_1 d\zeta_2^*, \tag{2.9}$$

whereas also the first-order pdf is required to define the related *autocovariance*,

$$C(\mathbf{r}_1, \mathbf{r}_2) \stackrel{\Delta}{=} R(\mathbf{r}_1, \mathbf{r}_2) - \mu(\mathbf{r}_1) \mu^*(\mathbf{r}_2) = \langle z(\mathbf{r}_1) z^*(\mathbf{r}_2) \rangle - \langle z(\mathbf{r}_1) \rangle \langle z^*(\mathbf{r}_2) \rangle, \tag{2.10}$$

variance,

$$\sigma^2(\mathbf{r}) \stackrel{\Delta}{=} C(\mathbf{r}, \mathbf{r}) = R(\mathbf{r}, \mathbf{r}) - \mu^2(\mathbf{r}) = \langle |z(\mathbf{r})|^2 \rangle - |\langle z(\mathbf{r}) \rangle|^2, \tag{2.11}$$

and *normalized-autocovariance function*,

$$\rho(\mathbf{r}_1, \mathbf{r}_2) \stackrel{\Delta}{=} \frac{C(\mathbf{r}_1, \mathbf{r}_2)}{\sigma(\mathbf{r}_1) \sigma(\mathbf{r}_2)}, \tag{2.12}$$

the latter being coincident to the *normalized-autocorrelation function* if a zero-mean stochastic process is in order.

Autocorrelation, autocovariance, and their normalized counterparts can be equivalently defined using as independent variables $\mathbf{r} = \mathbf{r}_1$ and the vector distance $\boldsymbol{\tau} = \mathbf{r}_1 - \mathbf{r}_2$.

The variance of the (zero-mean) surface increments at a given distance is referred to as the *structure function* $Q(\boldsymbol{\tau})$ of the process $z(\mathbf{r})$:

$$Q(\mathbf{r}_1, \mathbf{r}_2) \stackrel{\Delta}{=} \langle |z(\mathbf{r}_1) - z(\mathbf{r}_2)|^2 \rangle. \tag{2.13}$$

Equations (2.9) through (2.13) make reference to a complex stochastic process; if the process is real, as appropriate to natural surfaces, conjugate values do not apply. Generalization to complex random processes has been presented here because the above-mentioned formula can be applied both to the surface model and to the scattered field: as a matter of fact, surfaces represented by means of stochastic processes scatter electromagnetic fields that, in the phasor domain, are conveniently represented as complex random-vector processes.

2.2.3. Stochastic Processes: a Relevant Property

A stochastic process is *strict-sense stationary* (SSS), or simply *stationary*, if its statistical properties at any order are invariant to any shift in the domain where it is defined. Hence,

$$z(\mathbf{r}) \doteq z(\mathbf{r} + \bar{\mathbf{r}}), \forall \bar{\mathbf{r}} \tag{2.14}$$

where the symbol $\doteq$ means that the two members have the same statistics.

A stochastic process is *wide-sense stationary* (WSS) if its mean is constant:

$$\mu(\mathbf{r}) = \mu, \tag{2.15}$$

and its autocorrelation depends only on the vector distance, or *space lag*, $\mathbf{r}_1 - \mathbf{r}_2 \equiv \tau$

$$R(\mathbf{r}_1, \mathbf{r}_2) = R(\mathbf{r}_1 - \mathbf{r}_2) = R(\tau). \tag{2.16}$$

For a WSS process, quantities at the first member of Equations (2.7) and (2.11) are space independent, and quantities at the first member of Equations (2.9), (2.10), and (2.12) are space-lag dependent only.

According to Equation (2.12), the *correlation coefficient* of a WSS stochastic process is the surface-autocovariance function normalized to its value at zero distance:

$$\rho(\tau) \triangleq \frac{C(\tau)}{\sigma^2}. \tag{2.17}$$

Hence, for any zero-mean WSS stochastic process, the correlation coefficient is the normalized-autocorrelation function, and its maximum value, which is reached for $\tau = 0$, is unitary.

Examination of Equations (2.14) through (2.16) shows that strict-sense stationarity implies wide-sense stationarity; conversely, wide-sense stationarity is not sufficient for strict-sense stationarity.

For stationary stochastic processes, the structure function depends on τ and does not provide any further information with respect to the autocorrelation function: in fact, these are related by the following simple relation:

$$\begin{aligned} Q(\tau) &= \langle |z(\mathbf{r}_1) - z(\mathbf{r}_2)|^2 \rangle \\ &= \langle (z(\mathbf{r}_1) - z(\mathbf{r}_2))(z(\mathbf{r}_1) - z(\mathbf{r}_2))^* \rangle \\ &= 2R(0) - R(\tau) - R^*(\tau). \end{aligned} \quad (2.18)$$

In particular, for zero-mean real stationary stochastic processes, Equation (2.18) reduces to

$$Q(\tau) = 2\sigma^2 [1 - \rho(\tau)]. \quad (2.19)$$

A final overall comment to simplify the notation is due. Whenever confusion does not arise, in Equations (2.1) through (2.9), the independent random variable ζ may take the same symbol as the stochastic process, z: this is also done, starting from the next section, throughout this book.

2.3. Spectral Characterization of Stochastic Processes

In this section, fundamentals of the spectral representation of the stochastic processes are reported. Zero-mean, $\langle z(\mathbf{r}) \rangle = 0$, stochastic processes are considered here, this condition being easily met for stationary processes via a simple shift of the coordinate-reference system.

A first-order spectral description is provided by means of the *characteristic function* of a stochastic process, defined as the following statistical mean:

$$\langle \exp(-i\xi z) \rangle = \int_{-\infty}^{+\infty} p(z, \mathbf{r}) \exp(-i\xi z) \, dz, \quad (2.20)$$

thus coincident with the Fourier Transform (FT) of the first-order-process pdf. Use of the characteristic function turns out to be convenient whenever the description of the stochastic process is made via some relevant averages of it. As a matter of fact, the stochastic moments of the process are obtained as the MacLaurin series expansion of the characteristic function.

A second-order spectral description is in order. In the following equations, it is required that the stochastic processes are stationary: this is the

2.3. Spectral Characterization of Stochastic Processes

case in classical modeling of random rough surfaces and in classical evaluation of the electromagnetic field scattered by such surfaces. Conversely, nonstationary processes are considered in Chapter 3, because they are most appropriate to natural-surfaces fractal modeling.

We introduce $z(\mathbf{r}, q)$, a space-truncated version of the generic element of the ensemble $z(\mathbf{r})$, namely $z(\mathbf{r}, q)$, is coincident with $z(\mathbf{r})$ for values of each component of the independent space variable between $[-q, q]$, and is zero outside. Symbolically,

$$z(\mathbf{r}, q) \stackrel{\Delta}{=} z(\mathbf{r}) \text{rect}\left[\frac{\mathbf{r}}{2q}\right], \tag{2.21}$$

whose FT, $Z(\boldsymbol{\kappa}, q)$ is

$$Z(\boldsymbol{\kappa}, q) \stackrel{\Delta}{=} \int_{-\infty}^{\infty} z(\mathbf{r}, q) \exp(-i\boldsymbol{\kappa} \cdot \mathbf{r}) \, d\mathbf{r}. \tag{2.22}$$

The power P associate to the stochastic process $z(\mathbf{r})$ can be defined as

$$P \stackrel{\Delta}{=} \lim_{q \to \infty} \left(\frac{1}{2q}\right)^n \int_{-\infty}^{\infty} \left\langle |z(\mathbf{r}, q)|^2 \right\rangle d\mathbf{r}, \tag{2.23}$$

where n is the number of components of the independent space variables. Equation (2.23) represents, in the limit for $q \to \infty$, the n-dimensional space average of the process statistical mean square, and is the appropriate definition of the process power P, valid also for the nonstationary case.

Similarly, the *Power-Density Spectrum* (PDS), or *power spectrum*, $W(\boldsymbol{\kappa})$, can be defined as

$$W(\boldsymbol{\kappa}) \stackrel{\Delta}{=} \lim_{q \to \infty} \left(\frac{1}{2q}\right)^n \left\langle |Z(\boldsymbol{\kappa}, q)|^2 \right\rangle, \tag{2.24}$$

where Z is the FT of z, notation that is used from now on and applies to stationary as well as nonstationary stochastic processes.

The next step is to relate the power associate to the statistical process, Equation (2.23), to the power spectrum, Equation (2.24). This is readily accomplished by using the Parseval equality, which allows evaluating the *process energy* both in the space and in the transform domain as

$$\int_{-\infty}^{\infty} |z(\mathbf{r}, q)|^2 \, d\mathbf{r} = \left(\frac{1}{2\pi}\right)^n \int_{-\infty}^{\infty} |Z(\boldsymbol{\kappa}, q)|^2 \, d\boldsymbol{\kappa}. \tag{2.25}$$

Dividing each member of Equation (2.25) by $(2q)^n$, implementing the statistical mean and the limit for $q \to \infty$, it turns out that

$$\lim_{q \to \infty} \left\langle \left(\frac{1}{2q}\right)^n \int_{-\infty}^{\infty} |z(\mathbf{r}, q)|^2 \, d\mathbf{r} \right\rangle$$
$$= \lim_{q \to \infty} \left\langle \left(\frac{1}{2q}\right)^n \left(\frac{1}{2\pi}\right)^n \int_{-\infty}^{\infty} |Z(\boldsymbol{\kappa}, q)|^2 \, d\boldsymbol{\kappa} \right\rangle; \qquad (2.26)$$

the statistical mean can now be moved inside the integrals, limit and integral operations are exchanged in the second member, and Definitions (2.23) and (2.24) are employed to reach the final result that justifies Definition (2.24):

$$P = \left(\frac{1}{2\pi}\right)^n \int_{-\infty}^{\infty} W(\boldsymbol{\kappa}) d\boldsymbol{\kappa}. \qquad (2.27)$$

The power spectrum can be linked to the autocorrelation function. As a matter of fact, substituting Equation (2.22) in Definition (2.24) and considering Equation (2.21), it turns out that

$$W(\boldsymbol{\kappa}) = \lim_{q \to \infty} \left(\frac{1}{2q}\right)^n \left\langle \int_{-q}^{q} \int_{-q}^{q} z(\mathbf{r}_1) z^*(\mathbf{r}_2) \right.$$
$$\left. \exp\left[-i\boldsymbol{\kappa} \cdot (\mathbf{r}_1 - \mathbf{r}_2)\right] d\mathbf{r}_1 d\mathbf{r}_2 \right\rangle. \qquad (2.28)$$

Exchange of the statistical average with the integrals leads to

$$W(\boldsymbol{\kappa}) = \lim_{q \to \infty} \left(\frac{1}{2q}\right)^n \int_{-\infty}^{\infty} d\mathbf{r}_1 \int_{-\infty}^{\infty} R(\mathbf{r}_1, \mathbf{r}_2) \exp\left[-i\boldsymbol{\kappa} \cdot (\mathbf{r}_1 - \mathbf{r}_2)\right]$$
$$\text{rect}\left(\frac{\mathbf{r}_1}{2q}\right) \text{rect}\left(\frac{\mathbf{r}_2}{2q}\right) d\mathbf{r}_2. \qquad (2.29)$$

The following coordinate transformation, whose Jacobian has unitary modulus, is implemented:

$$\begin{cases} \mathbf{r}_1 = \mathbf{r} + \dfrac{\boldsymbol{\tau}}{2} \\ \mathbf{r}_2 = \mathbf{r} - \dfrac{\boldsymbol{\tau}}{2} \end{cases}. \qquad (2.30)$$

2.3. Spectral Characterization of Stochastic Processes

A simple geometric construction shows that

$$\text{rect}\left(\frac{\mathbf{r}_1}{2q}\right)\text{rect}\left(\frac{\mathbf{r}_2}{2q}\right) = \text{rect}\left(\frac{\mathbf{r}+\boldsymbol{\tau}/2}{2q}\right)\text{rect}\left(\frac{\mathbf{r}-\boldsymbol{\tau}/2}{2q}\right)$$

$$= \text{rect}\left(\frac{\boldsymbol{\tau}/2}{2q}\right)\text{rect}\left(\frac{\mathbf{r}}{2q-|\boldsymbol{\tau}|}\right). \quad (2.31)$$

Then, exchanging the order of integration and moving the limit inside the integral in $\boldsymbol{\tau}$, it turns out that

$$W(\boldsymbol{\kappa}) = \int_{-\infty}^{\infty} d\boldsymbol{\tau} \exp(-i\boldsymbol{\kappa}\cdot\boldsymbol{\tau}) \left[\lim_{q\to\infty}\left(\frac{1}{2q}\right)^n \text{rect}\left(\frac{\boldsymbol{\tau}}{4q}\right)\right.$$

$$\left. \times \int_{-\infty}^{\infty} R\left(\mathbf{r}+\frac{\boldsymbol{\tau}}{2},\mathbf{r}-\frac{\boldsymbol{\tau}}{2}\right)\text{rect}\left(\frac{\mathbf{r}}{2q-|\boldsymbol{\tau}|}\right)d\mathbf{r}\right]. \quad (2.32)$$

In this chapter, stationary surfaces are in order, so that $R\left(\mathbf{r}+\frac{\boldsymbol{\tau}}{2},\mathbf{r}-\frac{\boldsymbol{\tau}}{2}\right) = R(\boldsymbol{\tau})$. Hence,

$$W(\boldsymbol{\kappa}) = \int_{-\infty}^{\infty} d\boldsymbol{\tau} R(\boldsymbol{\tau}) \exp(-i\boldsymbol{\kappa}\cdot\boldsymbol{\tau})$$

$$\times \left[\lim_{q\to\infty}\left(\frac{1}{2q}\right)^n \text{rect}\left(\frac{\boldsymbol{\tau}}{4q}\right)\int_{-\infty}^{\infty}\text{rect}\left(\frac{\mathbf{r}}{2q-|\boldsymbol{\tau}|}\right)d\mathbf{r}\right]$$

$$= \int_{-\infty}^{\infty} d\boldsymbol{\tau} R(\boldsymbol{\tau}) \exp(-i\boldsymbol{\kappa}\cdot\boldsymbol{\tau})$$

$$\times \left[\lim_{q\to\infty}\left(\frac{1}{2q}\right)^n \text{rect}\left(\frac{\boldsymbol{\tau}}{4q}\right)(2q-|\boldsymbol{\tau}|)^n\right]. \quad (2.33)$$

Equation (2.33) is further simplified by implementing the limit $q \to \infty$. The rect $(\boldsymbol{\tau}/4q)$ states that each component of the variable $\boldsymbol{\tau}$ cannot exceed in module $2q$: when $q \to \infty$ it follows that rect $(\boldsymbol{\tau}/4q) \to 1$. In addition, $(2q - |\boldsymbol{\tau}|)^n$ can be asymptotically ($q \to \infty$) replaced by $(2q)^n$: as a matter of fact, the difference between these two quantities divided by $(2q)^n$ vanishes for $q \to \infty$. Thus, the power spectrum can be expressed as follows:

$$W(\boldsymbol{\kappa}) = \int_{-\infty}^{\infty} R(\boldsymbol{\tau})\exp(-i\boldsymbol{\kappa}\cdot\boldsymbol{\tau})d\boldsymbol{\tau}. \quad (2.34)$$

Equation (2.34) is referred to as the Wiener-Khinchin theorem, specialized to stationary processes: the power spectrum $W(\boldsymbol{\kappa})$ turns out to be the FT of the autocorrelation function. A slightly different formulation is obtained in Chapter 3 for the case of nonstationary processes.

The power spectrum provides an alternative popular representation for a stochastic process, and can be applied to describe both the scattering roughness and the scattered field. In the following chapters, it is shown that the surface-roughness power spectrum directly enters into the closed-form evaluation of the scattered field; more generally, the classical evaluation of the field scattered from stationary natural surfaces often requires the knowledge of the *generalized power-density spectrum*, $W^{(m)}(\kappa)$, defined as the FT of the autocorrelation function up to a certain power m:

$$W^{(m)}(\kappa) \triangleq \int_{-\infty}^{\infty} R^m(\tau) \exp(-i\kappa \cdot \tau) \, d\tau. \quad (2.35)$$

Obviously, $W^{(1)}(\kappa) = W(\kappa)$.

Equations (2.34) and (2.35) do not apply to nonstationary processes: in this case, the correlation function depends on two vector variables, and appropriate spatial averaging in Equations (2.34) and (2.35) is required (see Chapter 3).

2.4. Isotropic Surfaces

Whenever stochastic processes are used to describe a random rough surface, then the vector variable **r** can be expressed in terms of two independent scalar space variables—namely, x and y, when reference is made to a Cartesian coordinate-reference system. The same reference system, (O, x, y, z) (see Figure 2.1) is naturally used to analytically evaluate the scattered field.

A one-dimensional surface profile is obtained by intersecting the two-dimensional rough surface with a plane parallel to the z-axis. In particular, one-dimensional profiles shown in the figures of this chapter are obtained by intersecting the corresponding two-dimensional surfaces with the plane $y = 0$.

A random rough surface is said to be *isotropic* if the same stochastic characterization is held by all one-dimensional profiles crossing the origin O: hence, an isotropic surface is statistically invariant for any choice of the orientation of the Cartesian coordinate system (O, x, y).

For isotropic stationary surfaces, the surface description in the space and the spectral domain depends only on τ and κ, respectively, being $\tau = |\boldsymbol{\tau}|$

2.4. Isotropic Surfaces

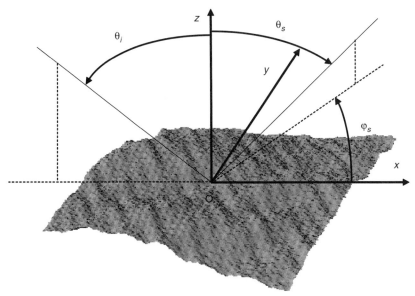

FIGURE 2.1 The Cartesian reference coordinate system.

and $\kappa = |\boldsymbol{\kappa}|$: thus, $R(\boldsymbol{\tau}) = R(\tau)$ and $W(\boldsymbol{\kappa}) = W(\kappa)$. Letting $\tau_x = (x_1 - x_2) = \tau \cos\varphi$, $\tau_y = (y_1 - y_2) = \tau \sin\varphi$, $k_x = \mathbf{k} \cdot \hat{\mathbf{x}} = k\cos\psi$, $k_y = \mathbf{k} \cdot \hat{\mathbf{y}} = k\sin\psi$, then

$$\tau = \sqrt{\tau_x^2 + \tau_y^2} = |\mathbf{r}_1 - \mathbf{r}_2| = \sqrt{(x_1 - x_2)^2 + (y_1 - y_2)^2}, \qquad (2.36)$$

$$\kappa = \sqrt{\kappa_x^2 + \kappa_y^2}, \qquad (2.37)$$

κ_x and κ_y being the Fourier mates of τ_x and τ_y respectively.

For isotropic stationary surfaces, Equation (2.35) can be rewritten as

$$\begin{aligned} W^{(m)}(\kappa) &= \int_{-\infty}^{\infty} R^m(\tau) \exp(-i\boldsymbol{\kappa} \cdot \boldsymbol{\tau}) d\boldsymbol{\tau} \\ &= \int_0^{\infty} R^m(\tau)\tau d\tau \int_0^{2\pi} \exp[-i\tau\kappa\cos(\varphi - \psi)] d\varphi \\ &= 2\pi \int_0^{\infty} J_0(\kappa\tau) R^m(\tau)\tau d\tau, \end{aligned} \qquad (2.38)$$

where use has been made of Equation (A.2.1), letting $z = \kappa\tau$ and $\nu = 0$. In particular, Equation (2.34) can be rewritten as

$$W(\kappa) = 2\pi \int_0^\infty J_0(\kappa\tau) R(\tau)\tau \, d\tau. \tag{2.39}$$

Equation (2.34) can be inverted accordingly to the Fourier Transform theory: repeating the same procedure leading to Equation (2.38), it turns out that

$$R(\tau) = \frac{1}{2\pi} \int_0^\infty J_0(\kappa\tau) W(\kappa)\kappa \, d\kappa. \tag{2.40}$$

Equations (2.39) and (2.40) allow us to move from the autocorrelation function to the power spectrum and vice versa, whenever isotropic stationary processes are in order.

For isotropic stationary processes, the structure function can be easily related to the isotropic surface spectrum: from Equations (2.18) and (2.40), it turns out that

$$Q(\tau) = 2\{R(0) - \text{Re}[R(\tau)]\}$$
$$= 2\left(\frac{1}{2\pi}\right)^2 \int_{-\infty}^{\infty}\int_{-\infty}^{\infty} [1 - \cos(\boldsymbol{\kappa} \cdot \boldsymbol{\tau})]W(\boldsymbol{\kappa})\,d\boldsymbol{\kappa}$$
$$= 2\left(\frac{1}{2\pi}\right)^2 \int_{-\infty}^{\infty}\int_{-\infty}^{\infty} \left\{1 - \frac{1}{2}[\exp(i\boldsymbol{\kappa}\cdot\boldsymbol{\tau}) + \exp(-i\boldsymbol{\kappa}\cdot\boldsymbol{\tau})]\right\}$$
$$\times W(\boldsymbol{\kappa})\,d\boldsymbol{\kappa}. \tag{2.41}$$

Applying Equation (A.2.1) three times—with $\nu = 0$ and $z = 0, -\kappa\tau, \kappa\tau$, respectively— the following relation is obtained:

$$Q(\tau) = \frac{1}{\pi}\int_0^\infty [1 - J_0(\kappa\tau)]W(\kappa)\kappa\,d\kappa. \tag{2.42}$$

Again, inverting Equation (2.42) as before, it turns out that

$$W(\kappa) = 2\int_0^\infty [1 - J_0(\kappa\tau)]Q(\tau)\,\tau d\tau. \tag{2.43}$$

It has to be noted that the structure function can be defined also for nonstationary processes with stationary increments (see Section 2.2 and Chapter 3). Of course, in this case, Equations (2.34), (2.35), and (2.40) do not hold, because $R = R(\mathbf{r}, \tau)$, whereas Equations (2.42) and (2.43) still make sense because $Q = Q(\tau)$.

2.5. Classical Models for Natural Surfaces: First-Order Stochastic Characterization

In this section, first-order stochastic processes that are usually employed for modeling natural surfaces in electromagnetic-scattering theory are reported. Here and in the following sections, these models are defined as *classical*, at variance with the more recently used *fractal* ones, which are referred to in Chapter 3. In the approximate analytic evaluations of the scattered field, these classical-surface models are generally used up to their characterization to the first and second order only. Isotropic surfaces are hereafter considered, unless otherwise specified.

A rough scattering surface $z(x, y)$ is usually modeled by a stationary, or WSS, process. The first-order pdf is usually selected to be a zero-mean σ^2 variance *Gaussian distribution* $z \stackrel{\Delta}{=} \mathrm{N}(0, \sigma^2)$:

$$p(z) = \frac{1}{\sqrt{2\pi}\sigma} \exp\left(-\frac{z^2}{2\sigma^2}\right). \tag{2.44}$$

Note that the mean of the first-order pdf can always be set equal to zero, as in Equation (2.44), by a simple shift, along z, of the coordinate system adopted in the analytic description of the surface and in the formulation of the scattered field. Hence, to the first-order characterization, the surface roughness is described by a single parameter: the *standard deviation* σ. For its physical counterpart, see Section 2.7.

The *characteristic function* of the Gaussian random process is easily obtained by using Equation (A.2.2):

$$\begin{aligned}\langle \exp(-i\xi z)\rangle &= \int_{-\infty}^{\infty} \frac{1}{\sqrt{2\pi}\sigma} \exp\left(-\frac{z^2}{2\sigma^2}\right) \exp(-i\xi z)\, dz \\ &= \exp\left(-\frac{1}{2}\sigma^2 \xi^2\right).\end{aligned} \tag{2.45}$$

Hence, Gaussian first-order pdfs lead to Gaussian characteristic functions whose variance is the inverse of the pdf variance.

Plots of the Gaussian probability-density function and corresponding characteristic function are shown in Figure 2.2; several values of the parameter σ^2 are considered.

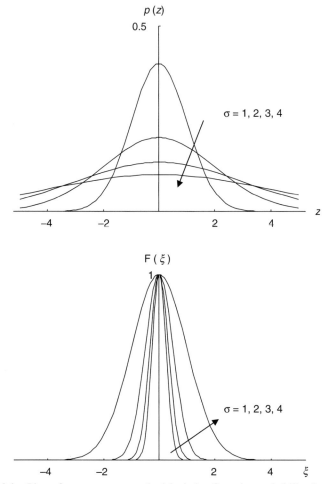

FIGURE 2.2 Plots of zero-mean σ-standard deviation Gaussian probability density functions $p(z)$ and corresponding characteristic functions $F(\xi)$. Values of $\sigma = 1, 2, 3, 4$ are considered.

2.6. Classical Models for Natural Surfaces: Second-Order Stochastic Characterization

The second-order pdf is usually assumed as jointly Gaussian. Considering a zero-mean isotropic surface, this joint pdf depends on a parameter σ and a function ρ:

$$p(z_1, z_2) = \frac{1}{2\pi\sigma^2\sqrt{1-\rho^2}} \exp\left(-\frac{z_1^2 - 2\rho z_1 z_2 + z_2^2}{2\sigma^2\left(1-\rho^2\right)}\right). \qquad (2.46)$$

The mathematical counterpart of the two parameters is immediate. Because

$$\int_{-\infty}^{\infty} p(z_1, z_2)\, dz_1 = \int_{-\infty}^{\infty} p(z_1, z_2)\, dz_2 = N\left(0, \sigma^2\right), \qquad (2.47)$$

then σ turns out to be the standard deviation of both the marginal zero-mean pdfs of z_1 and z_2. Moreover, substituting Equation (2.46) in Equation (2.9), it turns out that

$$R(z_1, z_2) = \sigma^2 \rho\left(\tau\right). \qquad (2.48)$$

Accordingly, ρ equals the normalized-autocorrelation function, defined for zero-mean stochastic processes by Equation (2.16), specified to the isotropic case. For the physical counterpart of the joint pdf parameters, see Section 2.7.

It has been noted in Section 2.5 that a single parameter, the standard deviation, is sufficient to characterize to the first order the zero-mean Gaussian stochastic process. Equation (2.46) shows that an additional function, the normalized-autocorrelation function, is needed for the second-order Gaussian characterization. An alternative characterization is possible, because the process is stationary: the power spectrum can be specified according to Equation (2.34). A further choice is the structure function, which is, for the considered stationary surfaces, simply related to the correlation coefficient and the power spectrum, see Equations (2.19) and (2.42).

The approach presented above systematically introduces the stochastic characterization of a stochastic process by considering subsequent orders for the joint pdfs; however, this is somehow redundant if only scattering computation from rough surfaces is in order. As a matter of fact, if the mean scattered field and scattered power must be evaluated, the classical-scattering approaches require only the knowledge of the surface first-order pdf and the surface autocorrelation (the latter being possibly replaced by the surface-structure function or power spectrum); hence, in those approaches, the second-order pdf is not required. In conclusion, classical evaluation of significant averages for the scattered stochastic field from a random surface usually requires prescription only of the first-order pdf and the surface autocorrelation. Equation (2.9) shows that the surface autocorrelation is obtained by performing an appropriate average over the second-order pdf; the latter cannot in general be specified in terms of the former; the Gaussian case—see Equations (2.46) and (2.48)—provides a noticeable exception to this rule. Then, in the following section, the first-order surface description is provided by means of a Gaussian pdf, whereas the second-order surface description is provided by prescribing only the autocorrelation function.

Several classical choices for this (partial) second-order characterization of a random rough surface to be employed within the classical scattering from rough surfaces are listed in Appendix 2.A; autocorrelation functions along with corresponding structure functions, power spectra, and generalized power spectra are collected in Tables 2.1 through 2.4.

Almost all reported autocorrelation functions share the important property that functions reported in Tables 2.2 through 2.4 can be analytically evaluated; moreover, power spectra and generalized power spectra enter directly into popular closed-form solutions for the mean scattered electromagnetic field and power density: it is concluded that use of surface models with the

TABLE 2.1 Some classical-surface models: normalized-autocorrelation functions. Isotropic surfaces are considered.

Stochastic processes	Normalized-autocorrelation function $\rho(\tau)$
Gaussian	$\exp\left(-\dfrac{\tau^2}{L^2}\right)$
Exponential	$\exp\left(-\dfrac{\tau}{L}\right)$
Intermediate Gaussian-exponential	$\exp\left[-\left(\dfrac{\tau}{L}\right)^\nu\right], \quad \nu \in [1, 2]$
Power-law	$\left[1 + \left(\dfrac{\tau}{L}\right)^2\right]^{-3/2}$
Multiscale Gaussian	$a\exp\left(-\dfrac{\tau^2}{L_1^2}\right) + b\exp\left(-\dfrac{\tau^2}{L_2^2}\right), \quad a+b=1$
Multiscale exponential	$a\exp\left(-\dfrac{\tau}{L_1}\right) + b\exp\left(-\dfrac{\tau}{L_2}\right), \quad a+b=1$
Mixed Gaussian-exponential	$a\exp\left(-\dfrac{\tau^2}{L_1^2}\right) + b\exp\left(-\dfrac{\tau}{L_2}\right), \quad a+b=1$

2.6. Second-Order Stochastic Characterization

TABLE 2.2 Some classical-surface models: structure functions. Isotropic surfaces are considered.

Stochastic processes	Structure function $Q(\tau)$
Gaussian	$2\sigma^2 \left[1 - \exp\left(-\frac{\tau^2}{L^2}\right)\right]$
Exponential	$2\sigma^2 \left[1 - \exp\left(-\frac{\tau}{L}\right)\right]$
Intermediate Gaussian-exponential	$2\sigma^2 \left\{1 - \exp\left[-\left(\frac{\tau}{L}\right)^\nu\right]\right\}, \quad \nu \in [1, 2]$
Power-law	$2\sigma^2 \left\{1 - \left[1 + \left(\frac{\tau}{L}\right)^2\right]^{-3/2}\right\}$
Multiscale Gaussian	$2\sigma^2 \left\{1 - \left[a\exp\left(-\frac{\tau^2}{L_1^2}\right) + b\exp\left(-\frac{\tau^2}{L_2^2}\right)\right]\right\}$, $a+b=1$
Multiscale exponential	$2\sigma^2 \left\{1 - \left[a\exp\left(-\frac{\tau}{L_1}\right) + b\exp\left(-\frac{\tau}{L_2}\right)\right]\right\}$, $a+b=1$
Mixed Gaussian-exponential	$2\sigma^2 \left\{\left[1 - \left[a\exp\left(-\frac{\tau^2}{L_1^2}\right) + b\exp\left(-\frac{\tau}{L_2}\right)\right]\right]\right\}$, $a+b=1$

autocorrelation functions reported in Appendix 2.A allow us to analytically evaluate in closed form the mean scattered field and power density. Hereafter, some more general comments on autocorrelation-function models are presented.

Other choices for the surface-autocorrelation function are obviously allowed, even if they do not lead to closed-form expression for the corresponding functions listed in Tables 2.2 through 2.4. However, any choice for $R(\tau)$ must satisfy properties held by real stochastic processes. For instance, the surface autocorrelation has to be an integrable real even function that reaches its maximum value in the origin and decreases monotonically for increasing distances τ.

TABLE 2.3 Some classical-surface models: power spectrum. Isotropic surfaces are considered.

Stochastic processes	Power spectrum $W(\kappa)$
Gaussian	$\pi \sigma^2 L^2 \exp\left[-\left(\dfrac{\kappa L}{2}\right)^2\right]$
Exponential	$2\pi \sigma^2 L^2 \left[1+(\kappa L)^2\right]^{-3/2}$
Intermediate Gaussian-exponential	no analytic result in closed form for general values of $\nu \in [1,2]$
Power-law	$2\pi \sigma^2 L^2 \exp(-\kappa L)$
Multiscale Gaussian	$\pi \sigma^2 L_1^2 \left\{ a\exp\left[-\left(\dfrac{\kappa L_1}{2}\right)^2\right] + b\left(\dfrac{L_2}{L_1}\right)^2 \exp\left[-\left(\dfrac{\kappa L_2}{2}\right)^2\right] \right\}$, $a+b=1$
Multiscale exponential	$2\pi \sigma^2 L_1^2 \left\{ a\left[1+(\kappa L_1)^2\right]^{-3/2} + b\left(\dfrac{L_2}{L_1}\right)^2 \left[1+(\kappa L_2)^2\right]^{-3/2} \right\}$, $a+b=1$
Mixed Gaussian-exponential	$\pi \sigma^2 L_1^2 \left\{ a\exp\left[-\left(\dfrac{\kappa L_1}{2}\right)^2\right] + 2b\left(\dfrac{L_2}{L_1}\right)^2 \left[1+(\kappa L_2)^2\right]^{-3/2} \right\}$, $a+b=1$

Let a zero-mean Gaussian first-order pdf be assumed: change of its standard deviation only acts as an overall amplification factor in the vertical scale. Similarly, once a choice has been made of the autocorrelation function, change of its main parameter, the correlation length, acts as an overall stretching factor in the horizontal scale. But also the shape of the autocorrelation function largely changes the surface correlation properties, hence, the surface shape. This behavior can be visually verified by inspection of Figures 2.4 through 2.10, presented in Appendix 2.A: surfaces with different autocorrelation shapes, but equal standard deviation and correlation length, look completely different.

The number of independent parameters appearing in the correlation function is another key issue to be dealt with in modeling the rough surface. Independent parameters are introduced to assure some model flexibility once the autocorrelation shape has been fixed: more specifically, they are

2.6. Second-Order Stochastic Characterization

TABLE 2.4 Some classical-surface models: generalized power spectra. Isotropic surfaces are considered.

Stochastic processes	Generalized power spectra $W^{(m)}(\kappa)$
Gaussian	$\dfrac{\pi \sigma^{2m} L^2}{m} \exp\left[-\dfrac{1}{m}\left(\dfrac{\kappa L}{2}\right)^2\right]$
Exponential	$2\pi \sigma^{2m} \left(\dfrac{L}{m}\right)^2 \left[1+\left(\dfrac{\kappa L}{m}\right)^2\right]^{-3/2}$
Intermediate Gaussian-exponential	no analytic result in closed form for general values of $1 < \nu < 2$
Power-law	$2\pi \sigma^{2m} \dfrac{L^2}{\Gamma\left(\dfrac{3}{2}m\right)} \left(\dfrac{\kappa L}{2}\right)^{\frac{3}{2}m-1} K_{1-\frac{3}{2}m}(\kappa L)$
Multiscale Gaussian	$\pi \sigma^{2m} L_1^2 L_2^2 \sum_{n=0}^{m} \binom{m}{n} \dfrac{a^{m-n} b^n}{n L_1^2 + (m-n) L_2^2}$ $\exp\left(-\dfrac{\kappa}{4} \dfrac{L_1^2 L_2^2}{n L_1^2 + (m-n) L_2^2}\right), \quad a+b=1$
Multiscale exponential	$2\pi \sigma^{2m} L_1^2 L_2^2 \sum_{n=0}^{m} \binom{m}{n} \dfrac{a^{m-n} b^n}{\left[n L_1 + (m-n) L_2\right]^2}$ $\left[1+\left(\dfrac{\kappa L_1 L_2}{n L_1 + (m-n) L_2}\right)^2\right]^{-3/2}, \quad a+b=1$
Mixed Gaussian-exponential	$2\pi \sigma^{2m} \left\{ a^m \left(\dfrac{L_1^2}{2m}\right) \exp\left[-\dfrac{1}{m}\left(\dfrac{\kappa L_1}{2}\right)^2\right] + b^m \left(\dfrac{L_2}{m}\right)^2 \left[1+\left(\dfrac{\kappa L_2}{m}\right)^2\right]^{-3/2} \right.$ $\left. + L_1^2 \sum_{n=1}^{m-1} \binom{m}{n} \dfrac{a^m b^{m-n}}{2n} \exp\left[-\dfrac{1}{n}\left(\dfrac{\kappa L_1}{2}\right)^2\right] \right\}, \quad a+b=1$

introduced to properly describe different correlation properties at different spatial scales that cannot be described by means of a single-parameter function. Hence, use of more than one independent parameter in the normalized-autocorrelation function often leads us to consider multiscale surfaces. A word of warning is now appropriate.

Models for the autocorrelation function or structure function or power spectrum, with increasing number of scales, certainly allow a better fit with experimental data: accordingly, use of two-scale autocorrelation, or structure function or power spectrum, is very popular when dealing with actual data. However, the analytic evaluation of the scattered field becomes cumbersome, and the interpretation of results questionable. In addition, it is worth arguing that the need of increasing number of scales is often a warning that the model may not be appropriate. Furthermore, any retrieval procedure of the surface parameters from backscattered data becomes almost not viable. This point is discussed in detail in Chapter 3, and it is one of the motivations for the introduction of fractal models in the natural-surfaces description.

2.7. Physical Counterpart of Natural-Surfaces Classical Parameters

In the preceding sections, the classical description of natural surfaces has been presented. It has been shown that some functions (pdfs or CDFs, along with their joint aggregate of any order) must be prescribed in order to fully describe the random surface. A simpler description is required if only some relevant averages of the scattered field must be evaluated: the classical choice is to postulate for the random surface a first-order description with a Gaussian behavior, and a partial second-order description via some allowed autocorrelation functions. In this case, the autocorrelation function is selected within a class involving the use of possibly few parameters; for stationary surfaces, those parameters are constant. In the simplest cases, only two parameters are necessary to describe the surface: a brief examination of Tables 2.1 and 2.2 shows that they are the *standard deviation* σ and the *correlation length* L.

The descriptions introduced in Section 2.6 and Appendix 2.A are rather simple; their use allows us to obtain an analytic solution for the scattered field, often expressed in closed form. Hence, those surfaces are usually employed in the scattering theory because they allow us to obtain simple analytic results; however, their use is highly questionable if a reliable representation of a natural scattering-surface is required. In summary, a classical-surface model very poorly (and often wrongly) describes a natural surface, but its scattered field can be evaluated in closed form.

These two parameters should obviously be *observable* quantities, thus either corresponding to or being related to physical entities. In this section, an attempt to elucidate this particular point is given. This is done for

2.7. Physical Counterpart of Natural-Surfaces Classical Parameters

σ and L separately. Results of this analysis are not only important per se, but also because they are a first step in trying to establish a link, under appropriate specific conditions, between classical and fractal parameters, the latter introduced in Chapter 3.

2.7.1. Standard Deviation

Seeking a *measure* of the deviation of the natural surface from a planar one is addressed. The attempt is to characterize the *roughness* of the surface.

The obvious candidate for this measure is the mean-square deviation of the surface height with respect to its mean plane. For the zero-mean Gaussian stationary isotropic case, it turns out that

$$\int_{-\infty}^{\infty} z^2 p(z)\, dz = \frac{1}{\sqrt{2\pi}\sigma} \int_{-\infty}^{\infty} z^2 \exp\left(-\frac{z^2}{2\sigma^2}\right) dz = \sigma^2, \quad (2.49)$$

as expected. Accordingly, an increase of the surface roughness leads to higher values of σ.

The probability that a point of the surface deviates from the plane more than σ is given by

$$\Pr\{|z(r)| \geq \sigma\} = 2 \int_{\sigma}^{\infty} p(z)\, dz \cong 0.317, \quad (2.50)$$

whereas $\Pr\{|z(r)| \geq 3\sigma\} \cong 0.003$. These results show not only that σ is an appropriate measure of the roughness, but also suggest a procedure (once this Gaussian pdf has been postulated) for its evaluation, starting from a large number of experiments. Alternatively, the power spectrum of the surface can be estimated and then evaluated for the wave number $\kappa \to 0$: examination of Table 2.3 shows that this value is proportional to σ^2 times the correlation parameter, or to a combination of the correlation lengths in the case of multiscale surfaces. This allows the evaluation of σ when the model has been postulated and the correlation parameter has been computed (see Section 2.7.2).

2.7.2. Correlation Length

Roughly speaking, the correlation length should be a measure of the constraint between height displacements of neighboring points of the surface: this constraint is expected to be significant if the points are well inside the correlation length and negligible outside it.

To further exploit this point, it is convenient to consider the jointly Gaussian pdf, Equation (2.46), and two points, z_1 and z_2, at a prescribed distance τ. The surface is assumed stationary and isotropic. Letting $z_2 = z_1 + \Delta z$, it turns out that

$$p(z_1, z_2) \to p(z_1, \Delta z)$$
$$= \frac{1}{2\pi\sigma^2\sqrt{1-\rho^2}} \exp\left[-\frac{2z_1^2(1-\rho) + 2z_1\Delta z(1-\rho) + (\Delta z)^2}{2\sigma^2(1-\rho^2)}\right], \tag{2.51}$$

where ρ is the normalized-autocorrelation-function coefficient computed at the distance τ. Integration over z_1 provides the pdf of the height change Δz:

$$p(\Delta z) = \int_{-\infty}^{\infty} p(z_1, \Delta z)\, dz_1 = \frac{1}{\sqrt{2\pi}\sqrt{2\sigma}\sqrt{1-\rho}} \exp\left[-\frac{(\Delta z)^2}{4\sigma^2(1-\rho)}\right]. \tag{2.52}$$

Results of Section 2.7.1 can now be applied: probability of excursions $|\Delta z|^2$ larger than few $2\sigma^2[1 - \rho(\tau)] = 2[\sigma^2 - R(\tau)]$ is marginal. Accordingly, if τ is small so that $R(\tau) \to \sigma^2$, then also the expected excursions Δz tend to zero. On the contrary, these oscillations can reach values of some σ if τ is large so that $R(\tau) \to 0$.

It is desirable to put the above qualitative statements on a more quantitative basis. The *effective correlation length*, L_e, of the autocorrelation function is now defined:

$$\int_0^{2\pi} d\vartheta \int_0^{\infty} R(\tau)\tau\, d\tau = R(0)\pi L_e^2, \tag{2.53}$$

with the scope of substituting to $R(\tau)$ an *effective* correlation function, constant and equal to its value $R(0)$ obtained at zero lag, inside the cylinder of radius L_e centered in the origin and zero outside (see Figure 2.3). For the Gaussian case $L_e = L$, see Table 2.1. Using this effective correlation function instead of the real one, the surface oscillations disappear for $\tau < L_e$, and are totally uncorrelated for $\tau > L_e$, being limited only by the mean-square-root deviation of the surface. It is therefore attractive to define L_e as the correlation length of the stochastic process.

This definition of the correlation length exhibits an additional interesting physical counterpart. As for the autocorrelation function, it is possible to

2.7. Physical Counterpart of Natural-Surfaces Classical Parameters

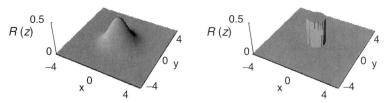

FIGURE 2.3 Equivalent correlation length. The Gaussian correlation function drawn on the left side holds the same correlation length of the equivalent cylindrical correlation function displayed on the right.

define an *effective spectral bandwidth*:

$$2\pi \int_{-\infty}^{\infty} W(\kappa)\kappa \, d\kappa = W(0)\pi \kappa_e^2. \tag{2.54}$$

By means of Equations (2.39) and (2.40), it turns out that

$$W(0) = 2\pi \int_0^{\infty} R(\tau)\tau \, d\tau = \pi R(0) L_e^2, \tag{2.55}$$

$$R(0) = \frac{1}{2\pi} \int_0^{\infty} W(\kappa)\kappa \, d\kappa = \frac{1}{(2\pi)^2} \pi W(0)\kappa_e^2. \tag{2.56}$$

Combining Equations (2.55) and (2.56), the link between effective correlation length and effective spectral bandwidth is obtained:

$$\kappa_e L_e = 2. \tag{2.57}$$

It follows that $\kappa_e/2\pi = 1/\pi L_e$ is a measure of the maximum spatial frequency exhibited by the randomly corrugated surface. It is concluded that L_e (equal to L in the Gaussian case) is an appropriate measure of the spatial stochastic *undulations* of the surface: if L_e decreases, the number of the surface undulations per unit length increases. In passim, the experimental estimate of the effective width of the power spectrum is a convenient way to determine L_e and L as well, once the surface model has been postulated.

As a last comment, note that a rough surface tends to reradiate an incoherent field when excited by a coherent electromagnetic wave. It is intuitive that the degree of incoherence increases if the roughness and undulations of the surface increase as well. Accordingly, a convenient measure of the difference between planar and rough surfaces is provided by the (adimensional) ratio

$$\Omega = \frac{\sigma}{L_e} = \frac{\kappa_e \sigma}{2}, \tag{2.58}$$

a parameter (σ/L in the Gaussian case) that can be referred to as the *incoherency parameter*.

2.8. Surface Classical Models Selection for Electromagnetic Scattering

In previous sections, it was shown that two independent functions must be assigned to fully describe the surface roughness up to the second order: in the stationary case, they are represented by the pdf $p(z)$ and the joint pdf $p(z_1, z_2)$.

If the first- and second-order pdfs are assumed to exhibit a Gaussian behavior, and a zero-mean isotropic stochastic process is in order, this is fully determined if the parameter σ and the autocorrelation function $R(\tau)$ are postulated. Alternatively, other functions may be used instead of $R(\tau)$: for instance, the structure function, which can be defined also for nonstationary surfaces, or the power spectrum $W(\kappa)$, which, in the stationary case, is related to $R(\tau)$ by a simple FT. These data fully describe this random surface up to the second order, in the Gaussian case; in general, the second-order description is not complete, but, in any case, this is sufficient to evaluate some relevant averages of the scattered electromagnetic field. It is important to note that at least the surface power spectrum in closed form is required by the classical methods to evaluate the mean and mean-square values of the scattered field in closed form.

More specifically, classical-scattering methods express the scattered field in terms of the surface standard deviation and the power spectrum, possibly generalized as reported in Table 2.4. Some of these spectral functions can be analytically evaluated in closed form, starting from the postulated autocorrelation function, as detailed in Appendix 2.A.

2.9. References and Further Readings

Fundamentals on random variables and stochastic process can be found in the work of Papoulis (1965). As far as their application within the electromagnetic theory is concerned, Ishimaru's study (1993) provides details on the structure function. Some of the classical functions presented in this chapter to describe random rough surfaces are also employed in

Fung's work (1994) within the integral-equation method to compute the scattered field.

Appendix 2.A   Surface Classical Models

For most applications, the characterization of stochastic surfaces is often provided up to the second order. This is particularly true within the electromagnetic scattering theory, where, in particular, Gaussian models are usually employed to describe the shape of the scattering surfaces: this corresponds to considering stationary isotropic surfaces, to select for the surfaces' shape a first-order Gaussian pdf, usually with zero-mean and prescribed variance σ^2, and a second-order Gaussian joint pdf with prescribed normalized autocorrelation $\rho(\tau)$. Possible choices for $\rho(\tau)$ are selected according to two requirements: first, to allow the analytic evaluation in closed form of the scattered field; and second, to comply with rough-surface modeling. Accordingly, several choices have been proposed and are reported in this appendix. It is worth recalling that these choices do not provide fully satisfactory models if natural surfaces and their scattered fields are in order: this being the reason to move to fractal models.

Gaussian, exponential, intermediate Gaussian-exponential, power-law, multiscale Gaussian, and multiscale exponential normalized-autocorrelation functions are considered in this appendix. For each choice, the corresponding structure function (using Equation [2.19]), power spectrum (using Equation [2.39]), and generalized power spectrum (using Equation [2.38]) are computed, possibly in closed form. The computed functions can be directly inserted into the classical electromagnetic-scattering formulations to get closed-form solutions; for instance, the surface power spectrum is required by the SPM classical solution, whereas also the generalized power spectrum is required by the Physical-Optics (PO) solution to the KA. According to those models, stationary isotropic surfaces are considered here. All results are referred to in Tables 2.1 through 2.4, and are illustrated by graphs depicted in Figures 2.4 through 2.10. Note that all graphs in this appendix are properly normalized. The horizontal scale for the autocorrelation function is normalized to L; for the structure function, the power spectrum, and the generalized power spectra, it is normalized to $1/L^2$ (to $1/L_1^2$ in the multiscale case). The vertical scale is normalized to σ^2 for the structure function; to $\sigma^2 L^2$, (to $\sigma^2 L_1^2$ in the multiscale case) for the power spectrum; and to $\sigma^{2m} L^2$ (to $\sigma^{2m} L_1^2$ in the multiscale case) for the generalized power spectra. When the normalized spectra are used to compute an element of the ensemble of the surface height $z(\mathbf{r})$, the vertical and horizontal scales are automatically normalized to σ and L (to L_1 in the multiscale case).

2.A.1. Gaussian Autocorrelation

The simplest choice for the surface normalized autocorrelation is a Gaussian function of prescribed *correlation length* $L > 0$:

$$\rho(\tau) = \exp\left(-\frac{\tau^2}{L^2}\right). \tag{2.A.1}$$

Using also Equation (A.2.2), it turns out that this choice implies an inverse Gaussian shape for the structure function, the power spectrum, and its generalization. Hence, the choice of Gaussian correlation simplifies most of the analytic evaluations involved in the scattered-field computation, and allows a closed-form solution for the scattered field.

Plots of the autocorrelation function, structure function, power spectrum, and generalized power spectra are shown in Figure 2.4a. A realization of this kind of rough surface is also plotted along with a surface cut in Figure 2.4b.

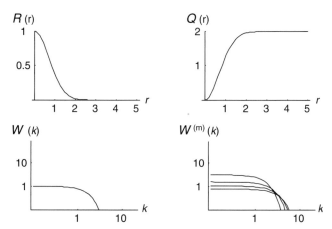

FIGURE 2.4a Plots of a Gaussian autocorrelation function and corresponding structure function, power spectrum and generalised power spectra of order $m = 1, 2, 3, 4$.

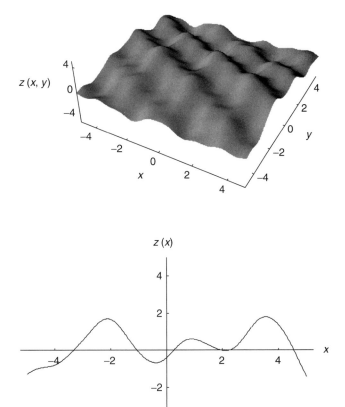

FIGURE 2.4b Graph of a realization of a rough surface with Gaussian autocorrelation and plot of the surface cut obtained by letting $y = 0$; the case $L = 1$ and $\sigma = 1$ is considered.

2.A.2. Exponential Autocorrelation

To model surfaces with higher correlation at larger distances, the normalized autocorrelation is chosen to be exponential of prescribed correlation length $L > 0$:

$$\rho(\tau) = \exp\left(-\frac{\tau}{L}\right). \tag{2.A.2}$$

This function is not differentiable at zero lag. This may pose some problems in the analytic evaluation of the scattered field, because the standard deviation of the derivative process turns out to be not defined.

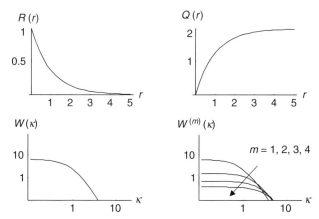

FIGURE 2.5a Plots of a normalized exponential autocorrelation function and corresponding structure function, power spectrum and generalized power spectra of order $m = 1, 2, 3, 4$.

The exponential choice implies an inverse exponential shape for the structure function. The power spectrum $W(\kappa)$, as well as its generalizations $W^{(m)}(\kappa)$, turns out to be of algebraic shape; only for $\kappa \gg 1/L$ is a power-law behavior recognized:

$$W(\kappa) \cong \frac{2\pi\sigma^2}{L}\kappa^{-3}, \tag{2.A.3}$$

and similarly for $W^{(m)}(\kappa)$.

As for the Gaussian choice, use of exponential autocorrelation is in favor of finding a closed-form solution for the scattered field.

Plots of the autocorrelation function, structure function, power spectrum, and generalized power spectra are shown in Figure 2.5a. A realization of this kind of rough surface is also plotted along with a surface cut in Figure 2.5b.

2.A.3. Intermediate Gaussian-Exponential Autocorrelation

A wide class of surfaces can also be represented by adopting an intermediate Gaussian-exponential autocorrelation function, the pure Gaussian and exponential autocorrelations being limiting cases:

$$\rho(\tau) = \exp\left[-\left(\frac{\tau}{L}\right)^\nu\right], \quad \nu \in [1, 2]. \tag{2.A.4}$$

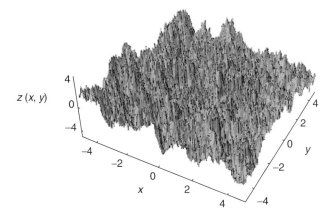

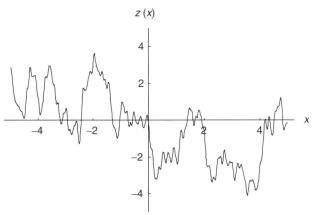

FIGURE 2.5b Graph of a realization of a rough surface with exponential autocorrelation and plot of the surface cut obtained by letting $y = 0$; the case $L = 1$ and $\sigma = 1$ is considered.

Correlation length, $L > 0$, and autocorrelation decay parameter, v, are usually prescribed to be independent; hence, this choice implies an intermediate inverse Gaussian-exponential shape for the structure function, as shown in Table 2.2. Power spectra cannot be evaluated in a closed form for any v. Hence, this choice of intermediate Gaussian-exponential autocorrelation is rarely adopted in electromagnetic applications, because it finds

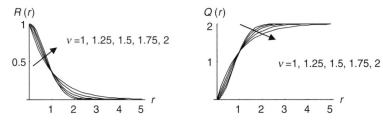

FIGURE 2.6 Plots of a normalized intermediate Gaussian-exponential autocorrelation function and corresponding structure function; the cases $v = 1, 1.25, 1.5, 1.75, 2$ are considered.

a closed-form solution for the scattered field in a very limited number of cases.

Plots of autocorrelation function and structure function are shown in Figure 2.6, parameterized to v.

2.A.4. Power-Law Autocorrelation

For isotropic surfaces, some power-law correlation functions turn out to be useful because their spectra can be evaluated in a closed form via a Bessel transform. For obtaining a power decay at large distance τ, the normalized autocorrelation may be set equal to

$$\rho(\tau) = \left[1 + \left(\frac{\tau}{L}\right)^2\right]^{-3/2}. \tag{2.A.5}$$

Because $\rho(L) \equiv 2^{-3/2} \cong 0.354 \cong 1/e$, the role played here by $L > 0$ is very similar to that exhibited in the Gaussian and exponential cases by the correlation length.

The structure function exhibits a powerlike shape as well, at least for distances much greater than the correlation length, as shown in Table 2.2.

The power spectrum exponentially decays with κ. A more complex shape, involving the gamma function $\Gamma(\cdot)$ and the Kelvin function of order v, $K_{-v}(\cdot)$, is exhibited by all the generalized spectra $W^{(m)}(\kappa)$ (see Table 2.4).

Choice of some power-law autocorrelations allows finding a closed-form solution for the scattered field.

Comparison of power-law and exponential choices shows that these two alternatives are dual each other: the power-law shape is met by the autocorrelation function of the former when τ/L is large, and by the power spectrum

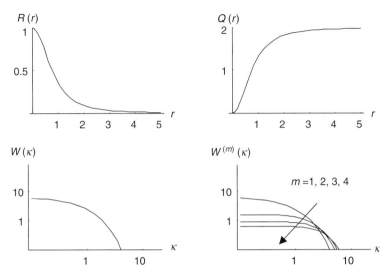

FIGURE 2.7a Plots of a normalized power-law autocorrelation function and corresponding structure function, power spectrum and generalized power spectra of order $m = 1, 2, 3, 4$.

of the latter when κL is large. In addition, the power-law autocorrelation is differentiable at the origin, at variance of the exponential one.

Plots of the autocorrelation function, structure function, power spectrum, and generalized power spectra are shown in Figure 2.7a. A realization of this kind of rough surface is also plotted along with a surface cut in Figure 2.7b.

2.A.5. Multiscale Gaussian Autocorrelation

To represent a wider class of surfaces, autocorrelation functions characterized by more than one parameter are introduced.

One of the simplest choices is the two-scale Gaussian autocorrelation function. For each term of the autocorrelation function, two correlation lengths, L_1 and L_2, are defined, and the normalized autocorrelation function is taken equal to

$$\rho(\tau) = a \exp\left(-\frac{\tau^2}{L_1^2}\right) + b \exp\left(-\frac{\tau^2}{L_2^2}\right), \quad a + b = 1. \quad (2.\text{A}.6)$$

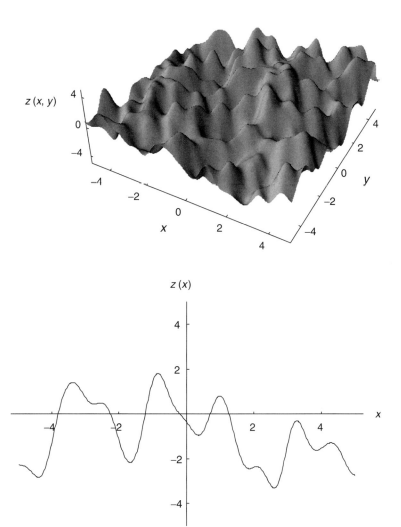

FIGURE 2.7b Graph of a realization of a rough surface with power-law autocorrelation and plot of the surface cut obtained by letting $y = 0$; the case $L = 1$ and $\sigma = 1$ is considered.

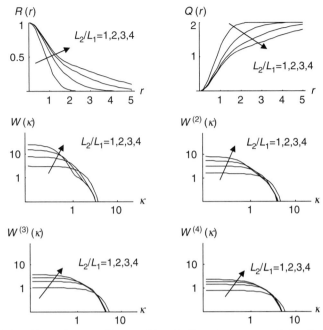

FIGURE 2.8a Plots of a normalized multi-scale Gaussian autocorrelation function and corresponding structure function, power spectrum and generalized power spectra of order $m = 2, 3, 4$; the cases $L_2/L_1 = 1, 2, 3, 4$, $a = b = 0.5$ are considered.

Hence, a two-scale Gaussian autocorrelation is a three-parameter function. Equation (2.A.6) implies a two-scale inverse Gaussian shape for the structure function and the power spectrum. Conversely, the generalized spectrum $W^{(m)}(\kappa)$ attains a more complicated shape: its expansion is given in Table 2.4.

The conclusion is that closed-form solutions for the scattered field are still possible, but the overall formulation becomes very much involved.

Extension to more than two scales can be easily provided for the autocorrelation, the structure function, and the power spectrum; generalized power spectra are less easy to obtain.

Plots of the autocorrelation function, structure function, power spectrum, and generalized power spectra are shown in Figure 2.8a; several values of L_1 and L_2 are considered, and all plots are parameterized to L_2/L_1. Realizations of this kind of rough surface are also plotted along with corresponding surface cuts in Figure 2.8b.

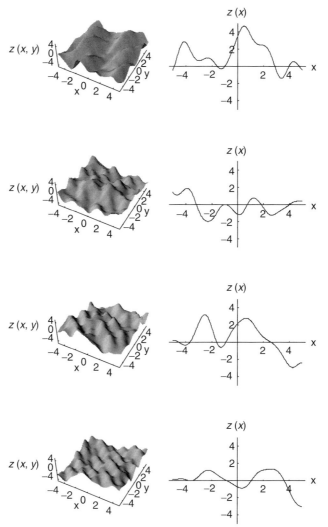

FIGURE 2.8b Graphs of realizations of rough surfaces with multi-scale Gaussian autocorrelation and plot of the surfaces cut obtained by letting $y = 0$; from top to down, the cases $L_2/L_1 = 1, 2, 3, 4$, $a = b = 0.5$ are considered.

2.A.6. Multiscale Exponential Autocorrelation

For multiscale surfaces exhibiting higher correlation for increasing distances compared to the multiscale Gaussian autocorrelation, the multiscale exponential is appropriate. A simple choice relies on a two-scale exponential autocorrelation function. Also in this case, two correlation lengths, L_1 and L_2, are defined, and the normalized autocorrelation function is taken equal to

$$\rho(\tau) = a \exp\left(-\frac{\tau}{L_1}\right) + b \exp\left(-\frac{\tau}{L_2}\right), \quad a+b=1. \quad (2.A.7)$$

Hence, the two-scale exponential autocorrelation is also a three-parameter function.

It is worth noting that, as in the case of exponential autocorrelation, this function is also not differentiable at zero lag. This may pose some problems in the analytic evaluation of the electromagnetic field scattered from such a surface, because the standard deviation of the derivative process is not defined.

The autocorrelation function in Equation (2.A.7) implies an inverse exponential shape for the structure function. For κ values much greater than $1/L_1$ and $1/L_2$, corresponding to the higher spatial frequencies (thus affecting the surface behavior at shortest distance spacings), the power spectrum turns out to be

$$W(\kappa) \cong 2\pi\sigma^2 \left(\frac{a}{L_1} + \frac{b}{L_2}\right) \kappa^{-3}, \quad (2.A.8)$$

thus exhibiting a power-law behavior. Conversely, a more complicated behavior is obtained for the generalized power spectrum.

As in the previous case, the conclusion is that closed-form solutions for the scattered field are still possible, but the overall formulation becomes very much involved.

Extension to more than two scales can be easily provided for the autocorrelation, the structure function, and the power spectrum; analytic evaluation of the generalized power spectra is not a straightforward task.

Plots of the autocorrelation function, structure function, power spectrum, and generalized power spectra are shown in Figure 2.9a; several values of L_1 and L_2 are considered, and all plots are parameterized to L_2/L_1. Realizations of this kind of rough surface are also plotted along with corresponding surface cuts in Figure 2.9b.

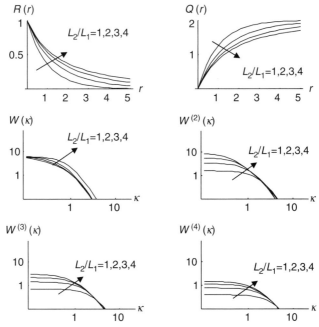

FIGURE 2.9a Plots of a normalized multi-scale exponential autocorrelation function and corresponding structure function, power spectrum and generalized power spectra of order $m = 2, 3, 4$; the cases $L_2/L_1 = 1, 2, 3, 4$, $a = b = 0.5$ are considered.

2.A.7. Mixed Gaussian-Exponential Autocorrelation

For multiscale autocorrelation models, it is also possible to prescribe a mixed behavior between Gaussian and exponential correlation, and the single-scale Gaussian and exponential autocorrelations can be recovered as limiting cases. Again, two correlation lengths, L_1 and L_2, are defined, and the normalized autocorrelation function is taken equal to

$$\rho(\tau) = a \exp\left(-\frac{\tau^2}{L_1^2}\right) + b \exp\left(-\frac{\tau}{L_2}\right), \quad a + b = 1. \quad (2.A.9)$$

This choice implies a mixed inverse Gaussian-exponential shape for the structure function, as well as for the power spectrum.

The more complicated form to the autocorrelation function implies that the generalized power spectra cannot be evaluated in a closed form for $m \neq 1$. An approximate expression is available, provided that the correlation length related to the Gaussian term is much lower than the correlation

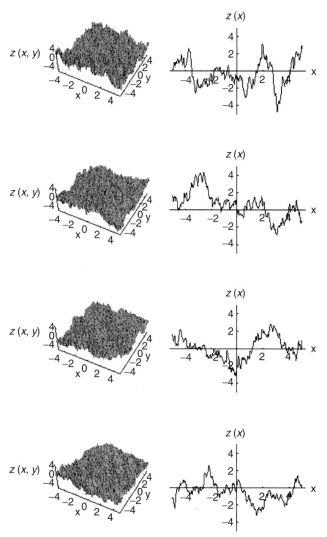

FIGURE 2.9b Graphs of realizations of rough surfaces with multi-scale exponential autocorrelation and plot of the surfaces cut obtained by letting $y = 0$; from top to bottom, the cases $L_2/L_1 = 1, 2, 3, 4$, $a = b = 0.5$ are considered.

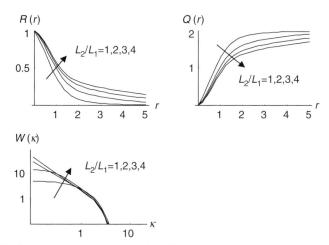

FIGURE 2.10a Plots of a normalized mixed Gaussian-exponential autocorrelation function and corresponding structure function and power spectrum; the cases $L_2/L_1 = 1, 2, 3, 4$, $a = b = 0.5$ are considered.

length related to the exponential term. Then, use of Equation (2.38) leads to the expression referred to in Table 2.4, and valid under the assumption $L_1 \ll L_2$.

It is concluded that the choice of a mixed Gaussian-exponential autocorrelation function allows developing some closed-form solutions for the scattered field.

Plots of the autocorrelation function, structure function, and power spectrum are shown in Figure 2.10a; several values of L_1 and L_2 are considered, and all plots are parameterized to L_2/L_1. Realizations of this kind of rough surface are also plotted along with corresponding surface cuts in Figure 2.10b.

Appendix 2.A Surface Classical Models

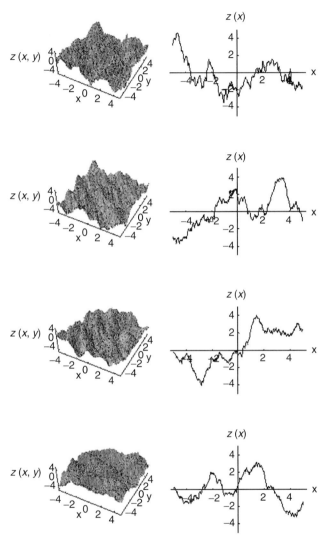

FIGURE 2.10b Graphs of realizations of rough surfaces with mixed Gaussian-exponential autocorrelation and plot of the surfaces cut obtained by letting $y = 0$; from top to bottom, the cases $L_2/L_1 = 1, 2, 3, 4$, $a = b = 0.5$ are considered.

CHAPTER 3

Surface Fractal Models

3.1. Introduction and Chapter Outline

In this chapter, the basic concepts of fractal geometry are presented. This is finalized to introduce symbols adopted in the book and to present main concepts and mathematical background relevant to model natural-surfaces roughness. The employed framework, in general, presents only main results from fractal geometry: mathematical details and results derivations are provided, if necessary, for dealing with the analytic evaluations of the electromagnetic field scattered from fractal surfaces; other elements of the theory are justified on intuitive bases. For exhaustive mathematical treatments and theorems proofs on fractal geometry, the reader may refer to the excellent mathematical books in this field listed under the References.

Strictly speaking, a unique definition of fractals does not exist. Mathematicians prefer to define fractals according to the properties of the fractal sets, just as biologists have no definition for life, and prefer to refer to the properties of living beings. Hence, fractal geometry is invoked for sets that hold a detailed structure on any arbitrary scale; are too irregular to be described according to classical geometry; hold some self-similarity or self-affine properties; are defined in very simple ways, often recursively; and hold a somehow defined fractal dimension larger than its topological one. These preliminary concepts are introduced and discussed in Section 3.2.

Sections 3.3 and 3.4 introduce two overall classification frameworks for fractal-surface models. Use of fractals in natural-surface modeling calls for *physical fractal surfaces*; their relationship with the corresponding *mathematical fractal surfaces* is presented and discussed in Section 3.3. Moreover, in Section 3.4, the rationale for distinction between *deterministic* and *stochastic fractal-surface* modeling is provided.

Sections 3.5 and 3.6 introduce the fractal models that can be employed to model natural-surfaces shape. Two fractal models are presented: the *fractional Brownian motion* (fBm) *process*, in Section 3.5, and the *Weierstrass-Mandelbrot* (WM) *function*, in Section 3.6. According to the classification introduced in Section 1.5, the former is a regular stochastic process, and the latter is a predictable process that allows us to model deterministic as well as random surfaces. In these sections, it is shown how to find appropriate models descriptions that can be used to express the electromagnetic scattered field in closed form: this is a crucial point, and great care is devoted to formally assessing it, thus providing a sound background of fractal-models use in the scattering problem. For instance, whenever surface spectral descriptions are introduced, their significance is rigorously assessed by making use of spectral theory for nonstationary processes.

Section 3.7 provides a connection between fBm processes and WM functions, thus showing that, independently from the mathematical employed model, the fractal surfaces hold a common rationale. Moreover, this connection is useful not only for comparing scattered-field results relying on the two fractal models, but also for providing the rationale for the synthesis of approximate fBm surfaces by using appropriately connected WM functions.

In Section 3.8, a fractal surface that is used as a reference in the following chapters to study the influence of the fractal parameters on the scattered electromagnetic field is introduced.

In Section 3.9, the existence of a link between classical and band-limited fractal parameters is discussed. Strictly speaking, there is no mathematical link between classical and fractal parameters: some fractal parameter, like the fractal dimension, has no counterpart in classical geometry, whereas some classical parameter, like the surface correlation length, is not uniquely defined in the fractal geometry, and in general diverges for mathematical fractal surfaces. However, a connection can be established whenever parameters relevant to band-limited fractals are in order; this connection may be useful in establishing a link between scattered-field results obtained by using classical and fractal geometries to model the natural surfaces.

The assumption that the fractals are more suitable than are conventional classical approaches to modeling natural surfaces is also confirmed by experiments and field campaigns, as referred to in Section 3.10. Moreover, natural-surface parameters estimation is in favor of fractal models. As a matter of fact, use of geometric models requires their parameters evaluation; in spite of the fact that classical-parameters evaluation relies on simple techniques, the values estimated on natural surfaces suffer of ambiguity and are not stable. Conversely, in spite of the fact that fractal-parameters evaluation may rely on involved techniques, the values estimated on natural surfaces are stable and not ambiguous. This is a further point in favor of the fractal description, which seems to have a perfect match with nature. And also it adds some value to the knowledge of the connection between the parameters that define the two descriptions.

Key references and suggestions for further readings are reported in Section 3.10.

3.2. Fundamentals of Fractal Sets

In this section, the fundamental concepts of fractal geometry as applied to describe natural surfaces are introduced. It is worth warning that some main concepts presented below are first rigorously stated, and then sometimes rephrased in an unconventional and definitively less rigorous style that might help the nonexpert reader to comprehend the overall framework.

The key parameter within fractal geometry is the fractal dimension: among several possible mathematical choices, the fractal dimension is defined in this book according to an appropriate measure. Fractal properties render the fractal models attractive and reliable to model natural surfaces; among fractal properties, the scaling ones are discussed in this section because they can be easily verified whenever natural surfaces are in order.

The fractal dimension employed throughout this book is the *Hausdorff*, or *Hausdorff-Besicovitch* (HB) *dimension*; its composite name is due to the fact that properties of the Hausdorff dimension are largely due to Besicovitch. The definition of the HB dimension is based on the definition of the Hausdorff measure. This measure is preliminarily introduced; it makes use of the Caratheodory method, consisting in defining measures as covers of sets.

3.2.1. Hausdorff Measure

Let U be any nonempty subset of a n-dimensional Euclidean space, R^n: the diameter $|U|$ of U is the greatest distance between two points belonging to U. A countable (or finite) collection of sets $\{U_i\}$ covers a set F if $F \subset \bigcup_i U_i$; furthermore, $\{U_i\}$ is said to be a δ-cover of F, if $0 < |U_i| \leq \delta$, $\forall i$.

The *s-dimensional Hausdorff measure* $H^s(F)$ of F is defined as

$$H^s(F) = \lim_{\delta \to 0} H^s_\delta(F), \quad \delta > 0, \ s > 0, \tag{3.1}$$

where

$$H^s_\delta(F) = \inf \left\{ \sum_{i=1}^\infty |U_i|^s : \{U_i\} \text{ is a } \delta\text{-cover of F} \right\}, \tag{3.2}$$

and $\inf\{\cdot\}$ is the lower bound. The superscript s in Equation (3.2) is a symbol in H^s_δ and an exponent in $|U_i|^s$.

In summary, Equation (3.2) deals with the sum of the s-th power of the diameters of δ-covering sets; the infimum of such a sum is found with respect to all permissible δ-covers of $F \in R^n$; as δ decreases, the class of permissible covers of F in Equation (3.2) reduces and $H^s_\delta(F)$ increases; the limit $\delta \to 0$ in Equation (3.1) is defined as the s-dimensional Hausdorff measure $H^s(F)$ of F; it exists for any set $F \in R^n$.

It can be shown that the Hausdorff measure generalizes, within a multiplicative constant, the Lebesgue measure as used, for instance, to define the number of points (R^0), the length (R^1), the area (R^2), and the volume (R^3) of Euclidean sets in R^n for $n = 0, 1, 2$, and 3, respectively.

It is convenient to close this subsection with a mathematically nonrigorous comment that can help in understanding the measure concept. Roughly speaking, for Euclidean sets, the diameter of each covering set U_i in Equation (3.2) is elevated to an integer n: for instance, for $n = 2$, the measure of F is obtained summing up the squared diameters (somehow the "areas") of the smallest sets that provide the best-fit covering of F. Then the Hausdorff measure appears to be consistent with intuition whenever measure of Euclidean sets is in order; moreover, it provides a formal extension that can be intuitively applied to measure irregular sets if, for any reason, they cannot be "fitted" in a collection of however small Euclidean covering sets.

3.2.2. Fractal Dimension

Once the Hausdorff measure has been presented, the Hausdorff dimension can be easily introduced. It is observed that the Hausdorff measure, as defined by means of the limit in Equation (3.1), diverges for s smaller than a certain threshold D, and equals zero for s greater than D. The critical value $s = D$ for which $H^s(F)$ changes from ∞ to 0 is defined as the HB *dimension* of F. In this book, the HB dimension of F is used as the *fractal dimension* for the surfaces under study.

The Hausdorff dimension D is not forced to be an integer number; if this is the case, $s = n$ is an integer and the Hausdorff measure generalizes, within a multiplicative constant, the Lebesgue measure. It appears that Euclidean dimensions are recovered when the Hausdorff measure coincides with integer values of D.

Intuitively speaking, any Euclidean set (point, curve, surface, volume, etc.) has Hausdorff measures that vary accordingly to the kind of covering sets: it is infinite if the covering is searched with sets of smaller dimensions, a real value if the covering is searched with sets of equal dimensions, zero if the covering is searched with sets of larger dimensions. For instance, this behavior is exhibited whenever segments, circles, and spheres, respectively, are used as covering sets for measuring the area of a (Euclidean) two-dimensional surface. Then, for a given set, the Hausdorff dimension is the only exponent to the covering sets' diameters that may provide a finite Hausdorff measure for the set.

The subset F of R^n, with $n > 1$, is *fractal* if its fractal dimension D is greater than its topological dimension, defined as equal to $n - 1$. Considering, for instance, the case $n = 3$, a surface in a three-dimensional space is fractal if its fractal dimension D is greater than 2 and smaller than 3. For a fractal surface, an infinite Hausdorff measure is obtained if one- or even two-dimensional covering sets are employed; conversely, a zero Hausdorff measure is obtained if three-dimensional covering sets are employed.

Before proceeding further along a rigorous mathematical treatment, a brief comment can be now appropriate to elucidate the above-mentioned concept of noninteger dimensions on a more intuitive but less rigorous basis. Roughly speaking, any fractal surface "fills" the space more than any however-complicated combination of Euclidean surfaces, but always leaves enough "unfilled" empty spaces whenever compared to any however-complicated combination of Euclidean volumes; and it is crucial to recall that these behaviors for fractal surfaces are true however "small" and whatever the shape of each element of the covering surfaces and volume. In other words, zooming in on a classical surface always leads to a sufficiently small spatial scale where the surface is revealed to be made joining regular (for instance, almost everywhere derivable), smooth, and possibly small Euclidean surfaces. Conversely, zooming in on a fractal surface reveals, at any however-small spatial scale, roughness details that cannot be constrained within a however-conceived collection of Euclidean surfaces, but these details never completely "fill" a however-shaped collection of Euclidean volumes. If the surface is represented via a mathematical function, then continuity is always verified, whereas derivability of such a mathematical fractal is never reached at any point and scale. In turn, the fractal dimension is a continuous parameter that quantifies the intermediate behavior held by any fractal surface between the two- and three-dimensional Euclidean limiting cases. And in particular, the HB dimension provides a sound mathematical background to define and apply this fractal-dimension concept.

Other definitions of the Hausdorff dimension, or finer definitions of dimensions, can be introduced (see Section 3.10); these are of some relevance whenever estimation of fractal parameters of a natural surface is in order, but they have no practical effect on the use of fractals in this book. Accordingly, this point will not be pursued.

For natural surfaces, the fractal dimension is directly related to the surface roughness in an intuitive way: an almost-smooth fractal surface has a fractal dimension slightly greater than 2, because is tends to flatten over a classically modeled surface; conversely, an extremely rough surface has a fractal dimension that approaches 3, because it tends to fill in a classical volume.

The fractal dimension is a fundamental parameter, but not the sole one specifying the fractal set: it is related to the fractal-set scaling properties, but does not prescribe an overall "amplitude" of the fractal set. In particular, for fractal surfaces, the fractal dimension, loosely speaking, states the relation among the different spectral components of the surface. It is evident that an additional independent overall spectral-amplification factor is needed to fully specify the surface spectrum and, consequently, the surface itself. This should not be surprising, and it is exploited in the subsequent sections: as a matter of fact, this is also true in the classical-surface description, where at least two parameters, the variance and the correlation length, are needed.

3.2.3. Scaling Properties

Among other interesting properties, the Hausdorff dimension allows us to verify the self-affine property that is common to fractal sets. This scaling property is also presented here because it provides a rationale to conceive fractal sets and distinguish them from Euclidean ones.

In the previous section, the HB dimension is introduced independently from the spatial scale: then fractal sets must exhibit some form of scale invariance to scaling transformations. The self-affine property quantifies such a scaling invariance, so that the surface roughness at different scales is related through a precise relationship.

Self-affinity generalizes the self-similarity property.

A set is *self-similar* if it is invariant with respect to any transformation in which all the coordinates are scaled by the same factor.

A set is *self-affine* if it is invariant with respect to any transformation in which the coordinates are scaled by factors that are in a prescribed relation; in such a case, the set fractal dimension is linked to the relation among the scaling factors.

Loosely speaking, for a self-similar set, a zooming function recovers scaled, and possibly rotated, copies of the original set: but for the scale, the original and scaled copies are indistinguishable from each other. This is shown in the example in the first and second row of Figure 3.1 for a generic set, and in Figure 3.2 for a surface. Conversely, for self-affine sets, the zooming function generates scaled, possibly rotated, and also appropriately stretched versions of the original set: this stretching can be appreciated by comparing the original and scaled copies. This is shown in the example in the first and third row of Figure 3.1 for a generic set, and in Figure 3.3 for a surface. Self-similarity and self-affinity hold in the statistical sense if invariance to corresponding coordinates scaling holds for the statistical distribution used to describe the set.

The s-dimensional Hausdorff measure (s is not forced to be an integer for fractal sets) scales with a factor γ^s if coordinates are scaled by γ. As a check, when s is an integer number and the Hausdorff measure generalizes the Lebesgue one, it is evident that the scaling property holds for lengths, areas, and volumes with s equal to 1, 2, and 3, respectively.

Note that the self-affine behavior is not sufficient to assess that a function—for example, the WM one—can represent a fractal, because other smooth nonfractal functions may exhibit this property and are not fractals. However, within some limitations, self-affinity may provide a rationale

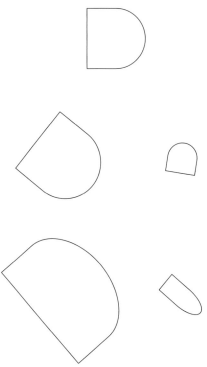

FIGURE 3.1 Self similarity and self-affinity: the starting set is depicted in the first row. Example of similarity transformations leads to the sets shown in the second row, whereas affine transformation results in the sets shown in the third row.

to conceive fractal sets and understand how they differ from the Euclidean ones. Intuitively speaking, if at some scale the fractal set "seems" to be covered by an appropriate collection of Euclidean sets, then it is sufficient to zoom in on the fractal set to uncover new details that "escape" from the previously considered cover; and this is true at any however-small scale. Hence, the fractal set cannot be measured by employing Euclidean covers, the Hausdorff measure is required, and the fractal dimension provides a norm to quantify the set behavior with respect to the zooming operation.

3.3. Mathematical versus Physical Fractal Sets

Self-affinity of fractal sets is the key mathematical property that makes them particularly useful for describing natural surfaces. However, whereas

3.3. Mathematical versus Physical Fractal Sets

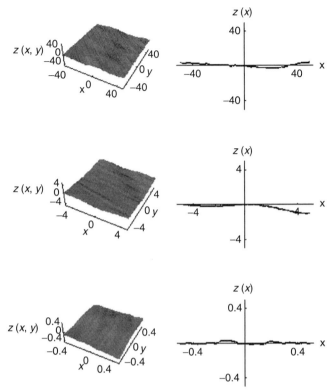

FIGURE 3.2 Self-similarity behavior of a rough surface. From top to bottom, zooming of a factor 10 along each axis is applied to successive surfaces and corresponding plots displaying the surface cut $y = 0$. The surface roughness appears statistically independent on the observation scale.

fractal sets maintain their self-affinity at any arbitrary observation scale, natural surfaces hold this relevant property only within inner- and outer-characteristic scales. In other words, natural surfaces exhibit fractal characteristics only on a possibly wide but limited range of scale lengths; these scale lengths represent the surface range of fractalness. Sets that exhibit an infinite range of fractalness are defined as *mathematical fractals*. Conversely, sets that exhibit a finite range of fractalness are defined as *physical fractals*.

In the following sections, it is shown how to define the spectral content of the fractal models. Observation scales correspond to surface-spectrum wavelengths: natural surfaces exhibit power-density spectra that are fractal only inside a finite bandwidth. This concept is further exploited if the

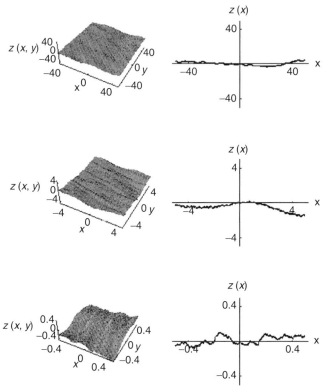

FIGURE 3.3 Self-affine behavior of a natural surface. From top to bottom, zooming of a factor 10 along each axis is applied between successive surfaces and corresponding plots displaying a generic vertical cut. The surface appears rougher whenever the observation scale is smaller.

observable natural surfaces are introduced—that is, if the description of the surface is related to a thought experiment.

Consider a surface to be explored: in particular, the height profile is of interest. Any actual instruments would sense, with a finite resolution, a limited length range of the observed surface. This can be equivalently stated assuming that the sensor is equipped with a band-pass filter: spatial scales and spectral bandwidth may be equally used to imply a finite exploration of the surface geometric properties. Each surface scale calls for a corresponding surface wave number; then range of fractalness calls for a corresponding fractalness bandwidth. It is concluded that a natural surface can be rigorously described by means of the corresponding mathematical fractal, provided that reference is made to range of scales, or bandwidth, within which the surface properties coincide with those of the mathematical fractal.

3.4. Deterministic versus Stochastic Fractal Description of Natural Surfaces

A natural surface can be represented by means of either a deterministic or a stochastic fractal set. The choice between these two very different representations relies on the applications to be made. In this section, the rationale to use a deterministic or a stochastic fractal description for natural surfaces is discussed.

Deterministic description of a rough-surface geometric shape may be accomplished by means of real fractal sets of topological (not fractal) dimension equal to 2. These sets may hold some relevant properties such as continuity, but may lack others, such as derivability. A relevant example is represented by the almost-periodic surfaces. In this case, the WM fractal function can be used to describe the surface roughness. Evaluation of the scattered field can then be performed by adopting this geometric description. In the following chapters, it is shown that this evaluation leads to closed-form solutions in terms of the geometric and electromagnetic parameters of the surface. It is important to underline that a closed-form solution for the field scattered from deterministic natural shapes can be obtained only by employing fractal models: as stated in Chapter 1, there is no classical deterministic description of natural surfaces that allows a closed-form solution for the scattered-field problem.

Random fractal surfaces can be described by means of randomized WM functions as well as of fBm stochastic processes. Randomized WM functions are predictable stochastic processes; conversely, fBm are regular stochastic processes. Then, accordingly with the considerations stated in Chapter 1, the WM functions allow evaluating the scattered field as a predictable process: the electromagnetic field is obtained as a function of some surface parameter. Conversely, the fBm model for the random surface allows evaluating the scattered field as a regular process: appropriate averages of the scattered-field distribution, usually the mean and the variance, can be computed as functions of the stochastic parameters of the surface probability distributions.

The WM function is a predictable process, so that it is possible to predict an entire WM sample function from its knowledge on a domain subset.

Conversely, the fBm process is a regular process, and it is possible to reconstruct only the statistics of the fBm random process from one sample function. These properties are fundamental for the selection of the type of the stochastic fractal model to be conveniently used for the evaluation of the scattered field. This applies also to the analysis and synthesis procedures of natural surfaces: the former consist of estimation of the fractal parameters from fractal-surface realizations; the latter are related to generation of fractal surfaces with prescribed fractal parameters.

3.5. Fractional Brownian Motion Process

In this section, the fBm is used to describe natural isotropic surfaces. Mathematical as well as physical fBm processes are discussed. To simplify the discussion of crucial and delicate points relevant to the use of this model, the particular case of isotropic surfaces is considered.

3.5.1. Mathematical fBm Processes
3.5.1.1. Definition

Definition of an fBm process is given in terms of the corresponding increment process. The stochastic process $z(x,y)$ describes an isotropic fBm surface if, for every x, y, x', y', all belonging to R, the increment process $z(x, y) - z(x', y')$ satisfies the following relation:

$$\Pr\{z(x,y) - z(x',y') < \bar{\zeta}\} = \frac{1}{\sqrt{2\pi} s \tau^H} \int_{-\infty}^{\bar{\zeta}} \exp\left(-\frac{\zeta^2}{2s^2 \tau^{2H}}\right) d\zeta, \quad (3.3)$$

$$\tau = \sqrt{(x-x')^2 + (y-y')^2},$$

where H is the *Hurst coefficient* (or *exponent*), and s is the standard deviation of surface increments at unitary distance measured in $[m^{(1-H)}]$. In the following discussion, to shorten notations, s and s^2 are referred to as the *surface incremental standard deviation* and *surface incremental variance*, respectively. In order to fully define an fBm process, the value of z at a given point should be specified: it is set $z(0) = 0$ so that, as shown below, the surface is self-affine. In the surface modeling and in the scattering problem, this condition is encompassed in the reference-system choice.

3.5. Fractional Brownian Motion Process

It can be demonstrated that a process satisfying Equation (3.3) exists if $0 < H < 1$, and that with probability 1, an fBm sample surface has a fractal (Hausdorff) dimension.

$$D = 3 - H. \tag{3.4}$$

Inspection of Equation (3.3) shows that if $\tau \to 0$, then $z(x,y) - z(x',y') \to 0$, thus proving that any sample function z is continuous with respect to x and y.

The incremental standard deviation s is related to a characteristic length of the fBm surface, the *topothesy*, T:

$$s = T^{(1-H)}. \tag{3.5}$$

By using Equation (3.5), Equation (3.3) can be written, in terms of topothesy and Hurst coefficient, in the equivalent form:

$$\Pr\{z(x,y) - z(x',y') < \bar{\zeta}\} = \frac{1}{\sqrt{2\pi} T^{(1-H)} \tau^H} \int_{-\infty}^{\bar{\zeta}} \exp\left(-\frac{\zeta^2}{2T^{2(1-H)} \tau^{2H}}\right) d\zeta. \tag{3.6}$$

Inspection of Equation (3.6) shows that the mean-square deviation of the surface increments equals $T^{(1-H)} \tau^H$. Furthermore, the surface-slope mean-square deviation is equal to the mean-square deviation of the surface increments divided by τ: $(T/\tau)^{1-H}$. It is concluded that topothesy is the distance (obviously measured in [m]) over which chords joining points on the surface have a surface-slope mean-square deviation equal to unity.

Some few relevant properties and crucial issues are presented in the following paragraphs for the fBm and the surface-increments processes.

3.5.1.2. *Order Characterization and Parameters Number*

Characterization of the surface to the first order is obtained from the increments-process definition by adding a reference value to the fBm in the origin $z(0,0) = 0$:

$$\Pr\{z(x,y) - z(0,0) < \bar{\zeta}\} = \Pr\{z(x,y) < \bar{\zeta}\}$$

$$= \frac{1}{\sqrt{2\pi} s \tau^H} \int_{-\infty}^{\bar{\zeta}} \exp\left(-\frac{\zeta^2}{2s^2 \tau^{2H}}\right) d\zeta,$$

$$\tau = \sqrt{x^2 + y^2}. \tag{3.7}$$

Examination of Equations (3.3) and (3.7) shows that the fBm increment process, as well as the first-order surface pdf, exhibits a Gaussian behavior.

Equation (3.3) specifies the pdf of the surface increments, $p(z_1 - z_2)$, which is not the same thing as the joint pdf $p(z_1, z_2)$. The surface-increments pdf is obtained from the joint pdf considering $z_2 = z_1 + \Delta z$ and integrating upon z_1, thus generalizing the procedure reported in Section 2.7.2 for the Gaussian second-order pdf. In general, the reverse is not true: the second-order pdf cannot always be fully deduced from knowledge of the first-order pdf of the increments process. Accordingly, up to this point, the second-order characterization of the fBm surface is not complete.

However, in subsequent chapters, it is shown that this partial second-order description of the fBm process is sufficient to represent natural surfaces, if only relevant averages of the scattered field—that is, its mean and power density—are of interest. Although a full stochastic characterization of the scattering surface is not achievable from the fBm fractal model represented by Definition (3.3) and the prescription $z(0) = 0$, the presented fractal model is sufficient to evaluate the scattered-power density from such a surface.

Then, as far as the number of parameters is concerned, the fBm process is a two-parameter regular random process. Definition of the fBm process has been done according to H and s in Equation (3.3), and to H and T in Equation (3.5). Other linked parameters, such as the fractal dimension D (see Equation [3.4]), can be used. The choice of the two independent parameters used to describe the process distribution is dictated by the use to be made of the stochastic process and by the mathematical- or physical-parameters interpretation.

Examples of fBm sample surfaces, and corresponding plots, are reported in the following discussion to show the influence of the fractal parameters on the surface appearance.

First, the effect of H is displayed by depicting in Figure 3.4 the fractal-surfaces graphs and corresponding plots for values of H ranging from 0.1 to 0.9, with step 0.2 and setting $T = 1$ m. These graphs show that the higher the H, the smoother the surface appears, thus visually confirming (see Equation [3.4]) a direct relationship existing between the fractal dimension D and an intuitive concept of "roughness." More specifically, Figure 3.4 and Equation (3.6) show that an increase of the fractal dimension does not correspond to a scaling of the graphs along the vertical axis: visual inspection of the figure shows that the increase of the fractal dimension generates surfaces whose closer spatial "undulations" become larger. Whenever the limit conditions $H \to 0$ and $H \to 1$ are considered, the

3.5. Fractional Brownian Motion Process

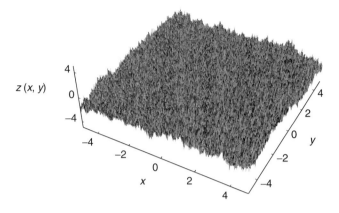

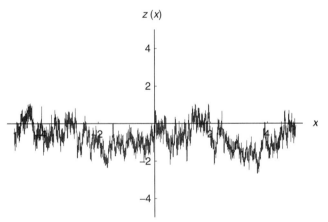

FIGURE 3.4a Fractal surfaces and corresponding profiles for H varying from 0.1 and 0.9, and a fixed value of $T = 1$ m. Case $H = 0.1$.

surface graphs and plots behaviors can be inferred as extrapolations of that reported in Figure 3.4. For $H \to 0$, the surface graphs tend to fill a 3-D volume, whereas for $H \to 1$, the surface tends to a classical one that is not flat if $T \neq 0$.

The dual case is analyzed by changing T and leaving H fixed to a constant value. In Figure 3.5, the cases corresponding to T ranging from 0.5 m to 1.5 m with step of 0.5 m and $H = 0.7$ are displayed. These graphs show

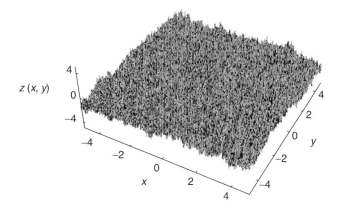

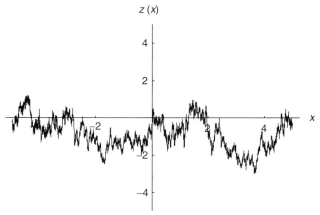

FIGURE 3.4b Fractal surfaces and corresponding profiles for H varying from 0.1 and 0.9, and a fixed valued of $T = 1$ m. Case $H = 0.3$.

that the higher T, the rougher the surface appears, thus visually confirming (see Equation [3.5]) a relationship existing between the topothesy T and an intuitive concept of "roughness." More specifically, visual inspection of the figure shows that an increase of the topothesy corresponds to a scaling of the graphs along the vertical axis, thus equally affecting all the space undulations.

Comparison of Figures 3.4 and 3.5 shows the different type of contributions provided by the two independent parameters D and T to the intuitive concept of "roughness": variation of the topothesy implies an overall scaling

3.5. Fractional Brownian Motion Process

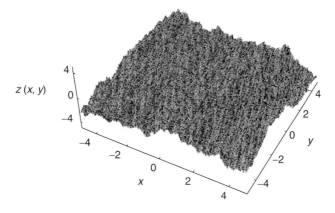

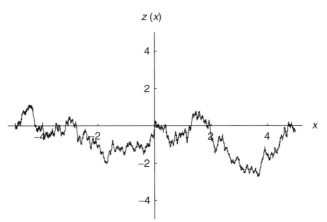

FIGURE 3.4c Fractal surfaces and corresponding profiles for H varying from 0.1 and 0.9, and a fixed valued of $T = 1$ m. Case $H = 0.5$.

to the surface roughness, whereas variation of the fractal dimension provides a different scaling for closer- and farther-surface "undulations."

In the following sections of this chapter, a quantitative rigorous mathematical support to this preliminary visual and intuitive analysis is provided.

3.5.1.3. Self-Affinity

As discussed in Section 3.2, a fundamental property of fractal sets is their self-affinity.

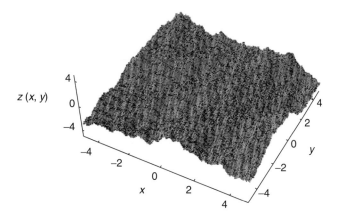

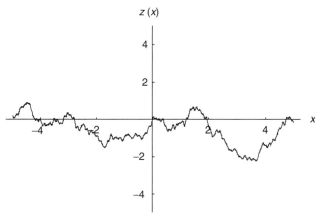

FIGURE 3.4d Fractal surfaces and corresponding profiles for H varying from 0.1 and 0.9, and a fixed valued of $T = 1$ m. Case $H = 0.7$.

By definition, fBm surfaces have increments that are self-affine in the statistical sense. Letting $\mathbf{r} = x\,\hat{\mathbf{x}} + y\,\hat{\mathbf{y}}$, $\mathbf{r}' = x\,\hat{\mathbf{x}}' + y\,\hat{\mathbf{y}}'$, $\Delta z(\tau) = z(\mathbf{r}) - z(\mathbf{r}')$, applying the definition of Equation (3.3) to $\Delta z(\lambda \tau)$, and changing the integration variable $\zeta/\gamma^H \to \zeta$, leads to

$$\Pr\left\{\Delta z(\gamma \tau) < \bar{\zeta}\right\} = \Pr\left\{\Delta z(\tau) < \frac{\bar{\zeta}}{\gamma^H}\right\}, \tag{3.8}$$

3.5. Fractional Brownian Motion Process

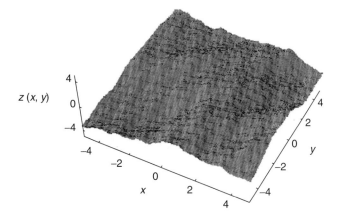

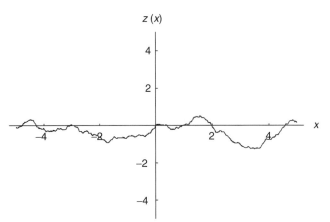

FIGURE 3.4e Fractal surfaces and corresponding profiles for H varying from 0.1 and 0.9, and a fixed valued of $T = 1$ m. Case $H = 0.9$.

for any $\gamma > 0$ and any τ and $\mathbf{r}$. It follows that

$$\Delta z(\gamma \tau) \doteq \gamma^H \Delta z(\tau), \qquad (3.9)$$

where, as in Chapter 2, the symbol $\doteq$ means "exhibits the same statistics as."

Self-affinity is also a property of the process z. As a matter of fact, letting $\mathbf{r}' = 0$ and $z(0) = 0$ in Equation (3.9), it turns out that for any $\gamma > 0$ and

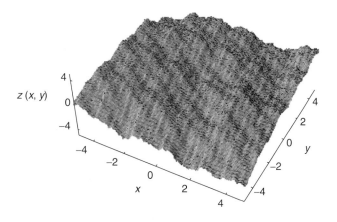

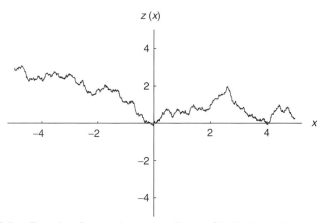

FIGURE 3.5a Fractal surfaces and corresponding profiles for T varying from 0.5 m to 1.5 m, and a fixed value of $H = 0.8$. Case $T = 0.5$ m.

for any τ,

$$z(\gamma\tau) \doteq \gamma^H z(\tau), \tag{3.10}$$

thus also proving that the isotropic process z is statistically self-affine.

3.5.1.4. Stationarity

Stationarity is a key property to be investigated for random processes. It is worthwhile verifying if this property is held by the fBm process.

3.5. Fractional Brownian Motion Process

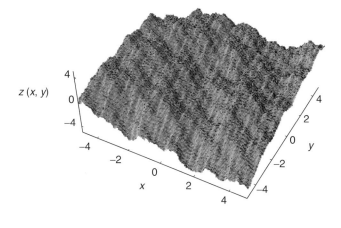

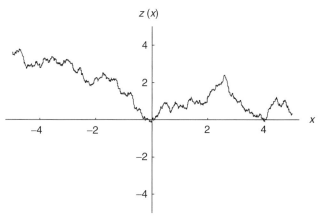

FIGURE 3.5b Fractal surfaces and corresponding profiles for T varying from 0.5 m to 1.5 m, and a fixed value of $H = 0.8$. Case $T = 1.0$ m.

The second member of Equation (3.3) represents the CDF of the increments of the $z(x, y)$ process. Hence, Equation (3.3) states that the height increments at any distance τ have a Gaussian distribution with zero mean and standard deviation $s\tau^H$. It is also immediately noted that the fBm $z(x, y)$ is a stochastic process with wide-sense stationary increments, because the second member of Equation (3.3) is dependent only on τ instead of x, y, x', y'.

However, stationarity is not a property held by the process z. As a matter of fact, taking into account that $z(0) = 0$, the autocorrelation of the fBm

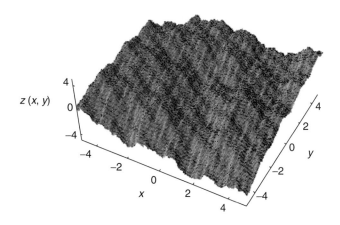

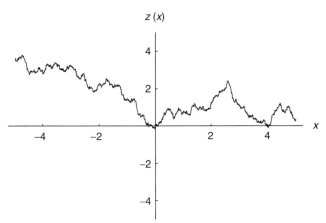

FIGURE 3.5c Fractal surfaces and corresponding profiles for T varying from 0.5 m to 1.5 m, and a fixed value of $H = 0.8$. Case $T = 1.5$ m.

process is computed as follows:

$$\begin{aligned}
R\left(\mathbf{r}, \mathbf{r}'\right) &= \left\langle z\left(\mathbf{r}\right) z\left(\mathbf{r}'\right)\right\rangle \\
&= \frac{1}{2}\left\langle z^2\left(\mathbf{r}\right) + z^2\left(\mathbf{r}'\right) - \left[z\left(\mathbf{r}\right) - z\left(\mathbf{r}'\right)\right]^2\right\rangle \\
&= \frac{1}{2}\left\langle \left[z\left(\mathbf{r}\right) - z\left(0\right)\right]^2 + \left[z\left(\mathbf{r}'\right) - z\left(0\right)\right]^2 - \left[z\left(\mathbf{r}\right) - z\left(\mathbf{r}'\right)\right]^2\right\rangle
\end{aligned}$$

(3.11)

3.5. Fractional Brownian Motion Process

$$= \frac{T^{2(1-H)}}{2}\left(|\mathbf{r}|^{2H} + |\mathbf{r}'|^{2H} - |\mathbf{r}' - \mathbf{r}|^{2H}\right),$$

because each quadratic factor has a Gaussian distribution (see Equation [3.6]), and its expected value coincides with its variance (see again Equation [3.6]). Hence, the fBm process is nonstationary because its autocorrelation is not dependent only on the space lag, $|\mathbf{r}' - \mathbf{r}|$, and particular attention must be paid in defining and evaluating its spectrum.

3.5.1.5. Structure Function

Natural random phenomena are frequently described by means of nonstationary stochastic processes holding the fundamental property to have stationary increments. For the particular case of nonstationary stochastic processes with stationary increments, it is particularly useful to compute the *structure function* $Q(\tau)$ of the process, defined as the variance of the (zero-mean) surface increments at given distance. In this relevant case, at variance to the process autocorrelation, this variance is a function only of the vector distance τ, and its relation with the power spectrum is simple to derive. Hence, the structure function for nonstationary processes with stationary increments represents a practical way to derive the power spectrum. This is discussed in the next sections.

According to Definitions (3.3) and (3.6), the fBm process of the increments over a fixed horizontal distance τ is a stationary isotropic zero-mean Gaussian process with variance equal to $s^2\tau^{2H} = T^{2(1-H)}\tau^2$:

$$\Delta z(\tau) \stackrel{\Delta}{=} N\left(0, s^2\tau^{2H}\right) = N\left(0, T^{2(1-H)}\tau^{2H}\right). \quad (3.12)$$

Therefore, evaluation of the structure function $Q(\tau)$ of an fBm process is straightforward, because, according to the definition given in Chapter 2, the structure function coincides with the variance of the process increments:

$$Q(\tau) = s^2\tau^{2H} = T^{2(1-H)}\tau^{2H}. \quad (3.13)$$

In Figure 3.6, the plot of the fBm normalized-structure function, parameterized to H, vs. the distance τ, shows that the variance of increments always increases with the distance τ; for *antipersistent* fBm ($0 < H < 1/2$), the increasing rate decreases with the distance—whereas for *persistent* fBm ($1/2 < H < 1$), it increases with the distance, the intermediate case corresponding to *Brownian motion* ($H = 1/2$) whose increments variance is proportional to the distance. The meanings of these two definitions, persistent and antipersistent, are explained in the following discussion. It is

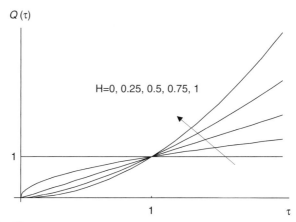

FIGURE 3.6 Plot of the fBm structure function for some relevant values of H. The horizontal axis is normalized to the topothesy T and the vertical axis to topothesy squared T^2.

convenient to evaluate the correlation of two successive increments over the surface,

$$\langle [z(\mathbf{r}) - z(\mathbf{r}')] [z(\mathbf{r}') - z(\mathbf{r}'')] \rangle$$
$$= \langle z(\mathbf{r}) z(\mathbf{r}') - z(\mathbf{r}) z(\mathbf{r}'') - z^2(\mathbf{r}') + z(\mathbf{r}') z(\mathbf{r}'') \rangle \qquad (3.14)$$
$$= \frac{T^{2(1-H)}}{2} \left(|\mathbf{r} - \mathbf{r}''| - |\mathbf{r} - \mathbf{r}'|^{2H} - |\mathbf{r}' - \mathbf{r}''|^{2H} \right),$$

wherein the latter expression has been obtained by means of the Equation (3.11) result. The correlation between successive increments in Equation (3.14) can assume positive as well as negative values. And "persistence" refers to the property of "saving" the feature of the increment along a generic path over the surface. If points $\mathbf{r}, \mathbf{r}', \mathbf{r}''$ are aligned along a generic line, the persistency property becomes transparent: in this case, $|\mathbf{r} - \mathbf{r}'| + |\mathbf{r}' - \mathbf{r}''| = |\mathbf{r} - \mathbf{r}''|$ and $|\mathbf{r} - \mathbf{r}'|^{2H} + |\mathbf{r}' - \mathbf{r}''|^{2H} > |\mathbf{r} - \mathbf{r}''|^{2H}$ if $2H < 1$, and vice versa, $|\mathbf{r} - \mathbf{r}'|^{2H} + |\mathbf{r}' - \mathbf{r}''|^{2H} < |\mathbf{r} - \mathbf{r}''|^{2H}$ if $2H > 1$.

It is concluded that antipersistent fBm ($0 < H < 1/2$) exhibits a negative correlation between successive increments whose projections onto the $z = 0$ plane are aligned; then, successive increments are more likely to have the opposite sign. Conversely, persistent fBm ($1/2 < H < 1$) exhibits positive correlation between successive increments whose projections onto the $z = 0$ plane are aligned; then, successive increments are more likely to have the same sign. Obviously, antipersistent and persistent behaviors emphasize whenever H diverges from $1/2$ and approaches 0 and 1, respectively. Finally, a Brownian motion ($H = 1/2$) has uncorrelated successive increments if their projections onto the $z = 0$ plane are aligned. Persistent and antipersistent behaviors relevant to fBm sample surfaces and plots are presented in Figure 3.7. Visual inspection of those figures and comparison with actual profiles suggests that natural surfaces more probably behave like persistent fBm stochastic processes.

3.5.1.6. *Power Spectrum*

In Chapter 2, it was shown that the power-density spectrum can be expressed in terms of the autocorrelation function as

$$W(\kappa) = \int_{-\infty}^{\infty} d\tau \exp(-i\kappa \cdot \tau) \left[\lim_{q \to \infty} \left(\frac{1}{2q} \right)^n \mathrm{rect}\left(\frac{\tau}{4q} \right) \right.$$
$$\left. \int_{-\infty}^{\infty} R\left(\mathbf{r} + \frac{\tau}{2}, \mathbf{r} - \frac{\tau}{2} \right) \mathrm{rect}\left(\frac{\mathbf{r}}{2q - |\tau|} \right) d\mathbf{r} \right]. \quad (3.15)$$

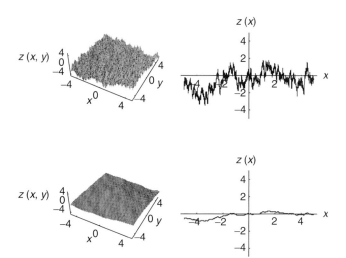

FIGURE 3.7 Fractal surfaces and corresponding profiles. First row: $H = 0.25$, corresponding to an anti-persistent fBm; second row: $H = 0.75$, corresponding to a persistent fBm.

This equation holds for stationary as well as nonstationary stochastic processes. In Chapter 2, it was also shown that in the case of stationary processes, the autocorrelation function does not depend on $\mathbf{r}$, and can be taken outside the inner integral in Equation (3.15), so that a simple FT relationship between the autocorrelation function and the power spectrum exists. For a nonstationary surface, the autocorrelation depends on $\mathbf{r}$, and different manipulations of Equation (3.15) are required to obtain the power-spectrum expression: this is done in the following discussion.

Equation (3.15) can be simplified by considering the limit $q \to \infty$. The rect($\tau/4q$) states that each component of the variable τ cannot exceed in module $2q$: when $q \to \infty$ it follows that rect($\tau/4q$) $\to 1$. In addition, the rect$[\mathbf{r}/(2q - |\tau|)]$ can be asymptotically ($q \to \infty$) replaced by rect$[\mathbf{r}/(2q)]$: as a matter of fact, this implies to extend, for each component, the integration interval of $R(\cdot)$ of a length $|\tau|$; being in that interval the autocorrelation limited, this extracontribution divided by $(2q)^n$ vanishes for $q \to \infty$. In conclusion:

$$W(\boldsymbol{\kappa}) = \int_{-\infty}^{\infty} d\boldsymbol{\tau} \exp(-i\boldsymbol{\kappa} \cdot \boldsymbol{\tau}) \left[\lim_{q \to \infty} \left(\frac{1}{2q} \right)^n \int_{-q}^{q} R\left(\mathbf{r} + \frac{\boldsymbol{\tau}}{2}, \mathbf{r} - \frac{\boldsymbol{\tau}}{2}\right) d\mathbf{r} \right]. \tag{3.16}$$

Thus, the power spectrum can be expressed as follows:

$$W(\boldsymbol{\kappa}) = \int_{-\infty}^{\infty} d\boldsymbol{\tau} \exp(-i\boldsymbol{\kappa} \cdot \boldsymbol{\tau}) \overline{R}(\boldsymbol{\tau}), \tag{3.17}$$

wherein

$$\overline{R}(\boldsymbol{\tau}) \triangleq \lim_{q \to \infty} \left(\frac{1}{2q} \right)^n \int_{-q}^{q} R\left(\mathbf{r} + \frac{\boldsymbol{\tau}}{2}, \mathbf{r} - \frac{\boldsymbol{\tau}}{2}\right) d\mathbf{r} \tag{3.18}$$

is the autocorrelation function averaged over the spatial variable $\mathbf{r}$.

Equations (3.17) and (3.18) are referred to as the Wiener-Khinchin theorem, referring to stationary as well as nonstationary processes, so that they generalize results obtained in Chapter 2. As a matter of fact, for a stationary process, $R = R(\boldsymbol{\tau})$ is taken outside the integral in $\mathbf{r}$ and $\overline{R}(\boldsymbol{\tau}) = R(\boldsymbol{\tau})$.

Substitution of Equation (3.11) in Equation (3.16) allows for evaluating the power spectrum of the fBm:

$$W(\boldsymbol{\kappa}) = \int_{-\infty}^{\infty} d\boldsymbol{\tau} \exp(-i\boldsymbol{\kappa} \cdot \boldsymbol{\tau}) \overline{R}(\boldsymbol{\tau})$$

3.5. Fractional Brownian Motion Process

$$= \int_{-\infty}^{\infty} d\tau \exp(-i\kappa \cdot \tau) \left[\lim_{q \to \infty} \left(\frac{1}{2q}\right)^2 \int_{-q}^{q} R\left(\mathbf{r} + \frac{\tau}{2}, \mathbf{r} - \frac{\tau}{2}\right) d\mathbf{r} \right]$$

$$= \int_{-\infty}^{\infty} d\tau \exp(-i\kappa \cdot \tau) \left[\lim_{q \to \infty} \left(\frac{1}{2q}\right)^2 \frac{T^{2(1-H)}}{2} \right.$$

$$\left. \int_{-q}^{q} \left(\left|\mathbf{r} + \frac{\tau}{2}\right|^{2H} + \left|\mathbf{r} - \frac{\tau}{2}\right|^{2H} - |\tau|^{2H} \right) d\mathbf{r} \right]. \quad (3.19)$$

Exchanging the integration limits, it turns out that

$$W(\kappa) = \lim_{q \to \infty} \left(\frac{1}{2q}\right)^2 \frac{T^{2(1-H)}}{2} \int_{-q}^{q} d\mathbf{r}$$

$$\int_{-\infty}^{\infty} d\tau \left(\left|\mathbf{r} + \frac{\tau}{2}\right|^{2H} + \left|\mathbf{r} - \frac{\tau}{2}\right|^{2H} - |\tau|^{2H} \right) \exp(-i\kappa \cdot \tau). \quad (3.20)$$

Computation of the FT of $|\tau|^{2H}$ requires resorting to generalized FTs (see Appendix 3.A):

$$\int_{-\infty}^{+\infty} |\tau|^{2H} \exp(-i\kappa \cdot \tau) d\tau = 2^{2+2H} \pi \frac{\Gamma(1+H)}{\Gamma(-H)} \frac{1}{\kappa^{2+2H}}$$

$$= -2^{2+2H} \Gamma^2(1+H) \sin(\pi H) \frac{1}{\kappa^{2+2H}}, \quad (3.21)$$

wherein Equations (A.3.2) and (A.3.3) have been applied to get the latter expression. Equation (3.21) exhibits a singularity in the origin. But, as explained in Appendix 3.A, the generalized FT is intended to be used after multiplication by a function of an appropriate set (essentially a filter), thus rendering it of physical interest. Application of the multiplication and shift rules to this generalized two-dimensional FT provides

$$\int_{-\infty}^{+\infty} \left|\frac{\tau}{2} \pm \mathbf{r}\right|^{2H} \exp(-i\kappa \cdot \tau) d\tau$$

$$= 2^2 \exp(\pm 2i\kappa \cdot \mathbf{r}) \int_{-\infty}^{+\infty} |\tau|^{2H} \exp(-i 2\kappa \cdot \tau) d\tau$$

$$= -2^2 \Gamma^2(1+H) \sin(\pi H) \frac{1}{\kappa^{2+2H}} \exp(\pm 2i\kappa \cdot \mathbf{r}). \quad (3.22)$$

Substitution of Equations (3.21) and (3.22) in Equation (3.20) and use of Relation (A.3.2) leads to the following result:

$$W(\kappa) = \frac{T^{2(1-H)}}{2} 2^{2+2H} \Gamma^2(1+H) \sin(\pi H) \frac{1}{\kappa^{2+2H}}$$

$$\lim_{q \to \infty} \left(\frac{1}{2q}\right)^2 \int_{-q}^{q} d\mathbf{r} \left[1 - 2^{1-2H} \cos(2\kappa \cdot \mathbf{r})\right]. \quad (3.23)$$

Evaluating the integral in Equation (3.23) leads to

$$W(\kappa) = \frac{T^{2(1-H)}}{2} 2^{2+2H} \Gamma^2(1+H) \sin(\pi H) \frac{1}{\kappa^{2+2H}}$$

$$\left[1 - 2^{1-2H} \lim_{q \to \infty} \left(\frac{1}{2q}\right)^2 \frac{\sin(2q\kappa_x)\sin(2q\kappa_y)}{\kappa_x \kappa_y}\right]. \quad (3.24)$$

In the limit $q \to \infty$, the sinusoidal term in brackets vanishes, and the power-density spectrum of the two-dimensional fBm exhibits an appropriate power-law behavior:

$$W(\kappa) = S_0 \kappa^{-\alpha}, \quad (3.25)$$

characterized by two spectral parameters—the *spectral amplitude*, S_0 measured in [m^{2-2H}], and the *spectral slope*, α—that, according to Equation (3.24), depend on the fractal parameters introduced in the space domain:

$$S_0 = 2^{2H+1} \Gamma^2(1+H) \sin(\pi H) T^{2(1-H)}$$
$$= 2^{2H+1} \Gamma^2(1+H) \sin(\pi H) s^2, \quad (3.26)$$

$$\alpha = 2 + 2H = 8 - 2D. \quad (3.27)$$

From the constraint on the Hurst exponent, $0 < H < 1$, it turns out that $2 < a < 4$, which defines the range of allowed values for the spectral slope α. Equations (3.26) and (3.27) state the relation between fractal parameters in the spectral domain, S_0 and α, and their mates H (or D) and T (or s) in the space domain.

The information content inherent to the six introduced fractal parameters and the mutual relationships between them deserves some comments.

Only two fBm parameters are independent. These two independent parameters can be selected, according to the corresponding constraints, in the spatial domain or, alternatively, in the spectral domain; formally,

3.5. Fractional Brownian Motion Process

TABLE 3.1 Synoptic view of the relationships between fractal parameters in the space and spectral domains for a topological two-dimensional surface. For any allowed choice of the fractal parameters reported in the first column, the remaining ones are evaluated in the second column.

H, s $0 < H < 1$	$D = 3 - H$ $T = s^{1/(1-H)}$ $\alpha = 2 + 2H$ $S_0 = 2^{2H+1}\Gamma^2(1+H)\sin(\pi H)s^2$
H, T $0 < H < 1$	$D = 3 - H$ $s = T^{1-H}$ $\alpha = 2 + 2H$ $S_0 = 2^{2H+1}\Gamma^2(1+H)\sin(\pi H)T^{2(1-H)}$
D, s $2 < D < 3$	$H = 3 - D$ $T = s^{1/(1-H)}$ $\alpha = 8 - 2D$ $S_0 = 2^{5-2D}\Gamma^2(4-D)\sin(\pi D)s^2$
D, T $2 < D < 3$	$H = 3 - D$ $s = T^{1-H}$ $\alpha = 8 - 2D$ $S_0 = 2^{5-2D}\Gamma^2(4-D)\sin(\pi D)T^{2(1-H)}$
S_0, α $2 < \alpha < 4$	$H = \alpha/2 - 1$ $D = 4 - \alpha/2$ $s = \{-S_0/[2^{\alpha-1}\Gamma^2(\alpha/2)\sin(\pi\alpha/2)]\}^{1/2}$ $T = \{-S_0/[2^{\alpha-1}\Gamma^2(\alpha/2)\sin(\pi\alpha/2)]\}^{1/(4-\alpha)}$

a mixed approach could be also followed, but the constraints could be difficult to interpret. A synoptic view of the relationships between fractals parameters in the space and in the spectral domain is reported in Table 3.1. Allowed independent parameters are always selected in pairs: one dimensionless parameter (H, D, or α) together with one dimensional parameter (T, s, S_0). For any pair of the independent fractal parameters, selected according to the corresponding constraint, the expressions for the remaining four ones are reported in Table 3.1.

Each one of the introduced pairs of fractal parameters describes the surface roughness from a different viewpoint. To exploit this point, it is convenient to refer to one possible choice, say the (H, s) pair. The variance of the surface increments, coincident with the structure function, is given by $Q(\tau) = s^2 \tau^{2H}$. Assuming that the surface roughness is related to the increments' variance, it depends on s and H, even if on a different footing. As a matter of fact, changes in the selected value of s imply that variations of the surface roughness are equal at any distance, whereas changing of H leads essentially to different variations of the surface roughness at different distances.

Once the spectral slope has been determined, a simple alternative evaluation of the spectral amplitude, S_0, Equation (3.26), can be obtained by employing the surface-structure function. As a matter of fact, substituting Equation (3.25) in Equation (2.42),

$$Q(\tau) = \frac{1}{\pi} \int_0^\infty [1 - J_0(\kappa\tau)] S_0 k^{-\alpha} \kappa \, d\kappa, \qquad (3.28)$$

making use of Equation (3.27) and integrating by parts, it turns out that

$$Q(\tau) = \frac{S_0 \tau}{2\pi H} \int_0^\infty J_1(\kappa\tau) \kappa^{-2H} \, d\kappa. \qquad (3.29)$$

Then, by using Equation (A.3.1) (with, $\mu = -2H$ and $\nu = 1$), the structure function can be evaluated in closed form:

$$\begin{aligned}Q(\tau) &= \frac{S_0}{2\pi H} 2^{-2H} \frac{\Gamma(1-H)}{\Gamma(1+H)} \tau^{2H} \\ &= \frac{S_0}{2^{2H+1} \Gamma^2(1+H) \sin(\pi H)} \tau^{2H}, \end{aligned} \qquad (3.30)$$

wherein Equations (A.3.2) and (A.3.3) have been taken into account. Comparison with Equation (3.13) leads to the required connection between the spectral slope and the fractal parameters in the space domain coincident with Equation (3.26).

It is concluded that the mathematical fBm process is not stationary, and suffers from the infinite-variance problem: in fact, being $\alpha \geq 2$, the integral of the fBm power spectrum diverges due to singularity in the low-frequency range (see Equation [3.25]). This low-frequency behavior of the spectrum is addressed as the *infrared catastrophe*. The reason is a historical one: the spectrum was examined in the optics area, where even surface wavelength of the order of microwaves may be addressed as infrared wavelengths.

Inspection of Equations (2.40) and (2.42) shows a fundamental result in the theory of nonstationary processes. Existence of the autocorrelation function requires that the power spectrum $W(\kappa)$ for $\kappa \to 0$ behaves like κ^α with $\alpha < 2$; conversely, existence of the structure function requires only $\alpha < 4$. If the process is stationary, the autocorrelation and the structure function both exist, whereas if the process is nonstationary with stationary increments, then the structure function does exist, and the autocorrelation, as a function of the space lag, does not. In the fBm fractal case $2 < a < 4$: then, the structure function always exists, whereas the autocorrelation function as a function of the space lag only, does not.

3.5.2. Physical fBm Processes

3.5.2.1. Definition

A surface satisfying Equation (3.3) for every τ is self-affine on all scales, so that it has details on any arbitrarily small scale: it is continuous, but not differentiable at any point. This poses problems when using Maxwell differential equations. For these reasons, together with those stated at the end of Section 3.5.1, *physical* fBm surfaces must be introduced. This is possible because no actual natural surface holds Property (3.3) at any scale, and some properties of fBm *mathematical* surfaces may be relaxed.

In Section 3.3, it was stated that natural surfaces exhibit a fractal behavior only on a wide but limited range of scales. A further limitation to the range of fractalness is imposed by the sensor applied to monitor the surface. The range of scales of interest for a scattering problem is limited on one side by the finite linear size l of the illuminated surface, or by the sensor resolution if processing of the received signal is implemented; and on the other side by the fact that surface variations on scales much smaller than the incident wavelength λ do not affect the scattered field. In most cases—in particular in remote sensing of natural surfaces at microwave frequencies—these limitations due to the employed sensor fall between the limits of intrinsic validity for the surface-shape fractal model. Accordingly, in electromagnetic scattering, *physical* fBm surfaces are considered; an efficient approach on surface modeling relies on considering surfaces that satisfy Equation (3.3) only for $\tau_m < \tau < \tau_M$, with τ_M of the order of l and τ_m usually taken of the order of $\lambda/10$. If $\tau_m \ll \tau_M$, then such surfaces satisfy Equation (3.25) only in a wide but limited range of spatial frequencies $\kappa_m < \kappa < \kappa_M$, with $\kappa_m \cong 1/\tau_m$. That is why these surfaces are also referred to as *band-limited* fBm, and for them the infrared catastrophe is avoided; they are stationary, at least in a wide sense, and regular.

3.5.2.2. Power Spectrum

It is shown hereafter that physical—that is, filtered fBm—fractals may be rendered stationary, at variance to fBm mathematical fractals.

A spatial filter function $g(\cdot)$ is applied to the fBm process: the filtered surface z^{filt} attains the following representation:

$$z^{filt}(\mathbf{r}) = \int_{-\infty}^{+\infty} z(\mathbf{t}) \, g(\mathbf{t} - \mathbf{r}) \, d\mathbf{t}, \quad (3.31)$$

wherein $g(\mathbf{r})$ is the inverse FT of the filter transfer function $G(\kappa)$. Integration is extended to the 2-D real plane. The filtered surface has zero mean, as follows by computing the statistical average of Equation (3.31): statistical average and integral operation are exchanged, and $\langle z^{filt}(\mathbf{t}) \rangle = 0$ because also the unfiltered fBm has zero mean upon setting $z(0) = 0$.

Evaluation of filtered surface autocorrelation is now in order:

$$R^{filt}(\mathbf{r}, \mathbf{r}') = \langle z^{filt}(\mathbf{r}) \, z^{filt}(\mathbf{r}') \rangle$$

$$= \frac{T^{2(1-H)}}{2} \int_{-\infty}^{+\infty} \int_{-\infty}^{+\infty} \left(|\mathbf{t}|^{2H} + |\mathbf{t}'|^{2H} - |\mathbf{t} - \mathbf{t}'|^{2H} \right) g(\mathbf{t} - \mathbf{r})$$

$$\times g(\mathbf{t}' - \mathbf{r}') \, d\mathbf{t} \, d\mathbf{t}', \quad (3.32)$$

where Equation (3.31) has been used, as well as statistical average, integrals exchange, and results of Equation (3.11) accounted for.

Equation (3.32) represents the autocorrelation function as the sum of three factors, each one expressed as the product of two integrals. The first two factors, after integration lead to a function which is not depending on $\mathbf{r}' - \mathbf{r}$ only: if it is required that the filtered process must be WSS, these factors must be made equal to zero, with an appropriate choice for the filter function.

The first factor, $R_1^{filt}(\mathbf{r}, \mathbf{r}')$, in Equation (3.32),

$$R_1^{filt}(\mathbf{r}, \mathbf{r}', a) = \frac{T^{2(1-H)}}{2} \int_{-\infty}^{+\infty} |\mathbf{t}|^{2H} g(\mathbf{t} - \mathbf{r}) \, d\mathbf{t} \int_{-\infty}^{+\infty} g(\mathbf{t}' - \mathbf{r}') \, d\mathbf{t}', \quad (3.33)$$

is considered first: appropriate sufficient conditions are searched to render it equal to zero.

Consider now a simple filter described by

$$\begin{cases} G(\kappa) = 1 & \forall \kappa \colon \kappa_m < \kappa < \kappa_M \\ G(0) = 0 & \text{elsewhere} \end{cases} \quad (3.34)$$

3.5. Fractional Brownian Motion Process

in the spectral domain. It turns out that

$$\int_{-\infty}^{+\infty} g\left(t'-r'\right) dt' = \int_{-\infty}^{+\infty} g\left(t'\right) dt' = G(0) = 0 \quad (3.35)$$

and Equation (3.33) is rendered equal to zero, provided that the other integral does not diverge. Then

$$\frac{T^{2(1-H)}}{2} \int_{-\infty}^{+\infty} |t|^{2H} g(t-r) dt = \frac{1}{(2\pi)^2} \int_{-\infty}^{+\infty} \frac{S_0}{\kappa^{2H+2}} G(\kappa) \exp(i\kappa \cdot r) d\kappa$$

$$= \frac{S_0}{2\pi} \int_{\kappa_m}^{\kappa_M} \frac{1}{\kappa^{2H+2}} J_0(\kappa r) \kappa d\kappa \neq \infty,$$

(3.36)

where Equations (3.21), (3.25), and (3.27) have been accounted for, and the usual transformation from Cartesian to polar integration coordinates implemented (see Equation [2.38]).

The same conclusion is reached for the second factor. Accordingly, the correlation function of the filtered surface reduces to the last third factor:

$$R^{filt}\left(r, r'\right) = \frac{T^{2(1-H)}}{2} \int_{-\infty}^{+\infty} |\xi|^{2H} d\xi \int_{-\infty}^{+\infty} g\left(t'-r'\right) g\left(t'-\xi-r\right) dt'$$

$$= \frac{T^{2(1-H)}}{2} \int_{-\infty}^{+\infty} |\xi|^{2H} d\xi \int_{-\infty}^{+\infty} g(\eta) g\left(\eta - \xi + r' - r\right) d\eta,$$

(3.37)

where the coordinate transformation $\xi = \left(t'-t\right)$, $\eta = \left(t'-r'\right)$ has been implemented. Equation (3.37) can be rewritten as

$$R^{filt}\left(r-r'\right) = \frac{T^{2(1-H)}}{2} \int_{-\infty}^{+\infty} |\xi|^{2H} R_g\left(\xi - r' - r\right) d\xi, \quad (3.38)$$

wherein

$$R_g(\xi) = \int_{-\infty}^{+\infty} g(\eta) g(\eta - \xi) d\eta$$

$$= \frac{1}{2\pi} \left[\kappa_M \frac{J_1(\kappa_M \xi)}{\kappa_M \xi} - \kappa_m \frac{J_1(\kappa_m \xi)}{\kappa_m \xi} \right]$$

(3.39)

is the autocorrelation of the filter function $g(\cdot)$.

It is concluded that the appropriately filtered fBm has zero mean, and its autocorrelation function depends on $r' - r$ only, so that it is stationary in the wide sense. The spectrum is now the FT of the correlation function, Equation (3.38), and is given by

$$W^{filt}(\kappa) = S_0 \kappa^{-\alpha} G(\kappa), \qquad (3.40)$$

where $G(\kappa)$ is given by Equation (3.34).

It must be remarked that the simple filter operation described by Equation (3.34) renders the surface process stationary. However, this filter with zero phase and sharp edges is not physically realizable: and the applied sensor is certainly characterized by a smoother filter. In this case, use of an equivalent effective spatial bandwidth, $\kappa_M - \kappa_m$, may be appropriate. In addition, the filter does not change the spectral properties of the fBm surface inside the spatial bandwidth of interest, but general features of the fBm process may be altered.

3.6. Weierstrass-Mandelbrot Function

In the following sections, WM functions used to describe natural surfaces are presented.

3.6.1. Mathematical WM Functions

Among several possible representations of the WM function, the most suitable one for modeling natural surfaces is a real function of two independent space variables x and y. A convenient choice is provided by the nonnormalized WM function $z(x,y)$, amenable to represent deterministic as well as random surfaces.

Consider the superposition of an infinite number of sinusoidal tones:

$$\begin{aligned} z(x,y) &= \sum_{p=-\infty}^{\infty} z_p \\ &= B \sum_{p=-\infty}^{\infty} C_p \nu^{-Hp} \sin\left[\kappa_0 \nu^p \left(x \cos \Psi_p + y \sin \Psi_p\right) + \Phi_p\right] \\ &= B \sum_{p=-\infty}^{\infty} C_p \nu^{-Hp} \sin\left[\boldsymbol{\kappa}_p \cdot \mathbf{r} + \Phi_p\right], \\ \boldsymbol{\kappa}_p &= \kappa_0 \nu^p \left(\hat{\mathbf{x}} \cos \Psi_p + \hat{\mathbf{y}} \sin \Psi_p\right), \end{aligned} \qquad (3.41)$$

3.6. Weierstrass-Mandelbrot Function

wherein $B[m]$ is the overall amplitude scaling factor; p is the tone index; $\kappa_0[m^{-1}]$ is the wavenumber of the fundamental component (corresponding to $p = 0$); $\nu > 1$ is the seed of the geometric progression that accounts for spectral separation of successive tones; $0 < H < 1$ is the Hurst exponent; and C_p, Ψ_p, Φ_p are deterministic or random coefficients that account for amplitude, direction, and phase of each tone, respectively.

Equation (3.41) exhibits a noninteger fractal dimension D as soon as ν is irrational, and the Hurst exponent is related to the fractal dimension $D = 3 - H$ as in Equation (3.4).

If the coefficients C_ps are deterministic, they must be all equal and constant: $C_p = C$, so that the tone amplitudes, $BC\nu^{-Hp}$, deterministically follow the power-law spectral behavior typical of fractal functions. For random coefficients C_ps, the usual choice for their pdf is Gaussian with zero mean and unitary variance.

If the coefficients Ψ_ps are deterministic, all equal and constant, $\Psi_p = \Psi$, the surface exhibits the fractal behavior only in the direction selected by Ψ and is constant along the direction orthogonal to it. If the coefficients Ψ_ps are uniformly distributed in $[-\pi, \pi)$, the WM function is isotropic in the statistical sense; any other choice leads to an anisotropic surface.

If the coefficients Φ_p s are deterministic, any choice assuring that the WM function deterministically exhibits the self-affine behavior is allowed. If the coefficients Φ_ps are random, they are usually chosen uniformly distributed in $[-\pi, \pi)$, and the *zero set* of the WM function—that is, the set of points of intersection with the plane $z = 0$—is nondeterministic.

In the case of a random WM function, the random coefficients, C_p, Ψ_p, Φ_p, are usually assumed to be mutually independent.

The WM function holds the self-affine behavior only for the discrete values of $\gamma = \nu^n$ unless $\nu \to 1$: in this case, the WM function approaches the self-affine behavior for every scaling factor γ. Accordingly, in the deterministic case,

$$z(\gamma x, \gamma y) = \gamma^H z(x, y), \qquad (3.42)$$

for $\nu \to 1$, and the WM function and its appropriately scaled versions are equal. Similarly, in the random case,

$$z(\gamma x, \gamma y) \doteq \gamma^H z(x, y) \qquad (3.43)$$

and the WM function and its appropriately scaled versions are statistically identical.

It is concluded that the self-affine behavior, typical of fractals including fBm, is strictly observed by the mathematical WM fractal function if $\nu \to 1$, irrespective of the random or deterministic behavior of the WM coefficients.

Computation of the power spectrum of the WM function is now in order.

Examination of Equation (3.41) suggests that the power spectrum is composed of lines centred at $\kappa = \kappa_p$, in view of the independence of C_ps coefficients:

$$W(\kappa) = \sum_{p=-\infty}^{\infty} P_p(\kappa_p) \frac{\delta(\kappa - \kappa_p)}{2\pi \kappa}, \qquad (3.44)$$

$P_p(\kappa)$ being the power associated to each tone. This, in turn, can be computed as

$$P_p(\kappa_p) = \langle z_p^2 \rangle = B^2 \langle C_p^2 \rangle \nu^{-2Hp} \langle \sin^2[\kappa_p \cdot \mathbf{r} + \Phi_p] \rangle = B^2 \frac{\nu^{-2Hp}}{2}, \quad (3.45)$$

upon averaging the C_p and Φ_p. Accordingly, the spectral lines of $W(\kappa)$ are given by

$$W(\kappa) = \frac{B^2}{2} \sum_{p=-\infty}^{\infty} \nu^{-2Hp} \frac{\delta(\kappa - \kappa_p)}{2\pi \kappa}, \qquad (3.46)$$

and cluster around the origin for large negative values of p, because κ_p is proportional to ν^p; the spectrum diverges with its integral there, so that the infrared catastrophe is present also for the WM function.

3.6.2. Physical WM Functions

Equation (3.41) shows that the mathematical WM fractal function is determined by the parameters B, H, ν, κ_0. Physical WM functions can be obtained just limiting the summation to P tones, thus obtaining band-limited WM surfaces:

$$z(x, y) = B \sum_{p=0}^{P-1} C_p \nu^{-Hp} \sin[\kappa_0 \nu^p (x \cos \Psi_p + y \sin \Psi_p) + \Phi_p]. \quad (3.47)$$

As in the case of fBm, use of band-limited WM surfaces is physically justified by the fact that surface fractality is held on a wide but limited range of scales, and any scattering measurement is limited to a finite set of scales. Let (X, Y) be the antenna footprint over the surface. The lowest spatial frequency

3.6. Weierstrass-Mandelbrot Function

of the surface, $\kappa_0/2\pi$, is linked to the footprint diameter $\sqrt{X^2+Y^2}$, possibly through an appropriate safety factor $\chi_1 \in (0,1]$, whereas its upper spatial frequency $\kappa_0 \nu^{P-1}/2\pi$ is related to the electromagnetic wavelength λ, possibly through an appropriate safety factor $\chi_2 \in (0,1]$, usually set equal to 0.1. Accordingly, we can set

$$\kappa_m = \kappa_0 = \frac{2\pi \chi_1}{\sqrt{X^2+Y^2}}, \qquad (3.48)$$

and

$$\kappa_M = \kappa_0 \nu^{(P-1)} = \frac{2\pi}{\chi_2 \lambda}. \qquad (3.49)$$

Definitions (3.48) and (3.49) can be combined to provide the number of tones, $P \in \mathbb{N}$, in terms of the sensor wavelength and footprint:

$$P = \left\lceil \frac{\ln\left(\sqrt{X^2+Y^2}\big/\chi_1\chi_2\lambda\right)}{\ln \nu} \right\rceil + 1, \qquad (3.50)$$

where $\lceil \cdot \rceil$ stands for the ceiling function, defined so as to take the upper integer of its argument.

It can be easily checked that the band-limited WM Surface (3.47) is stationary. The correlation function is given by

$$R(\mathbf{r}_1,\mathbf{r}_2) = \langle z(\mathbf{r}_1)z(\mathbf{r}_2) \rangle$$

$$= B^2 \sum_{p=0}^{P-1}\sum_{q=0}^{P-1} \langle C_p C_q \nu^{-H(p+q)} \sin[\boldsymbol{\kappa}_p \cdot \mathbf{r}_1 + \Phi_p] \sin[\boldsymbol{\kappa}_p \cdot \mathbf{r}_2 + \Phi_q] \rangle$$

$$= \frac{B^2}{2}\sum_{p=0}^{P-1}\sum_{q=0}^{P-1} \nu^{-H(p+q)} \langle C_p C_q \rangle \{\cos[\boldsymbol{\kappa}_p \cdot (\mathbf{r}_2-\mathbf{r}_1)+(\Phi_q-\Phi_p)]$$

$$+ \cos[\boldsymbol{\kappa}_p \cdot (\mathbf{r}_2+\mathbf{r}_1)+(\Phi_q+\Phi_p)]\},$$

$$= \frac{B^2}{2}\sum_{p=0}^{P-1} \nu^{-2Hp} \langle \cos[\boldsymbol{\kappa}_p \cdot (\mathbf{r}_2-\mathbf{r}_1)] \rangle$$

$$= R(\mathbf{r}_2-\mathbf{r}_1), \qquad (3.51)$$

where averaging has been implemented with respect to C_p and Φ_p.

Equation (3.51) can be further manipulated. Let $\mathbf{r}_2 - \mathbf{r}_1 = \tau \cos\varphi \hat{\mathbf{x}} - \tau \sin\varphi \hat{\mathbf{y}}$; then

$$\cos[\mathbf{\kappa}_p \cdot (\mathbf{r}_2 - \mathbf{r}_1)] = \Re\{\exp[i\kappa_p \tau \cos(\varphi - \psi_p)]\}$$

$$= \Re\left\{\sum_{n=-\infty}^{+\infty} J_n(\kappa_p \tau) \exp(in\pi) \exp[-in\cos(\varphi - \psi_p)]\right\}, \quad (3.52)$$

(see Equation [A.5.1]). Implementing the averaging process with respect to Ψ_p and substituting in Equation (3.51), the final result

$$R(\tau) = \frac{B^2}{2} \sum_{p=0}^{P-1} v^{-2Hp} J_0(\kappa_p \tau) \quad (3.53)$$

is obtained.

The power-density spectrum of physical WM functions is obtained by truncating Series (3.46) to the finite values of p belonging to the interval $[0, P-1]$.

3.7. Connection between fBm and WM Models

The connection between these two fractal models is appropriately found by comparing the spectral behavior of these two stochastic processes. The two representations cannot be equivalent in any respect because the independent parameters characterizing the fBm are two, for instance H and S_0, whereas in the WM, four independent parameters are considered, B, H, ν, κ_0. However, it is shown here that from the spectral point of view, the physical WM process is an appropriately sampled version of a band-limited fBm one.

Let the spectral plane κ_x, κ_y to be subdivided into concentric annular regions of radii $(\kappa_p \nu^{1/2}, \kappa_p \nu^{-1/2})$, respectively; then the spectral power within each annular region can be computed for the WM and fBm density-power spectra (addressed in the following discussion as spectra).

For the WM function, only the p-th tone is present in the annular region of central radius κ_p, hence:

$$P_{WM}(\kappa_p) = \frac{B^2}{2} v^{-2Hp} = \frac{B^2}{2\kappa_0^{-2H}} \kappa_p^{-2H}. \quad (3.54)$$

3.7. Connection between fBm and WM Models

For the fBm process, upon integration within the same annular spectral interval, it turns out that

$$
\begin{aligned}
P_{fBm}(\kappa_p) &= \left(\frac{1}{2\pi}\right)^2 \int_{\kappa_0 \nu^{p-1/2}}^{\kappa_0 \nu^{p+1/2}} 2\pi S_0 \kappa^{-(2H+1)} d\kappa \\
&= \frac{S_0}{4\pi H} \kappa_0^{-2H} \nu^{-2Hp} \left(\nu^H - \nu^{-H}\right) \\
&= \frac{S_0}{4\pi H} \left(\nu^H - \nu^{-H}\right) \kappa_p^{-2H}.
\end{aligned}
\tag{3.55}
$$

The spectral decays exhibited by Equations (3.54) and (3.55) coincide, provided that the Hurst exponent H is the same for the WM and fBm functions. If this is the case, then it is allowed to enforce the connection between the two spectra by equating the two powers, thus getting

$$
B^2 = \frac{S_0}{2\pi H} \kappa_0^{-2H} \left(\nu^H - \nu^{-H}\right).
\tag{3.56}
$$

If Condition (3.56) is verified, the WM spectrum is recognized to be a discretized version of the fBm spectrum at the discrete wave numbers $\kappa_p, p = 0, 1, \ldots$: each sample is representative of an appropriately defined annular region of the fBm spectrum.

The transition from the discrete WM to the continuous fBm spectrum is accomplished by recalling that the power is obtained by integration of the power-density spectrum divided by $(2\pi)^2$, according to the Parseval theorem. Equation (3.54) represents the power provided by the p-tone of the WM distribution within the spectrum annular region. Therefore, the equivalent WM power-density spectrum within the annular region is given by $(2\pi)^2$ times Equation (3.54) divided by the area of the annular region:

$$
(2\pi)^2 \frac{B^2}{2\kappa_0^{-2H}} \kappa_p^{-2H} \frac{1}{\pi \kappa_p^2 \left(\nu^{1/2} - \nu^{-1/2}\right)^2} = \frac{S_0}{H} \frac{\left(\nu^H - \nu^{-H}\right)}{\left(\nu^{1/2} - \nu^{-1/2}\right)^2} \kappa_p^{-(2H+2)},
\tag{3.57}
$$

where Equation (3.56) has been accounted for. Now the annular width $\kappa_p \left(\nu^{1/2} - \nu^{-1/2}\right)$ approaches zero, if $\nu \to 1$. In this limit, k_p is made a continuous variable, $\kappa_p \to \kappa$, and Relation (3.25) under Condition (3.56) is obtained, because the ratio of the two ν functions approaches H in the limit $\nu \to 1$.

It is concluded that whenever $\nu \to 1$, the spacing between successive tones of the WM functions tends to vanish, and the spectrum tends to become continuous, closely approximating that of the fBm process.

The above results can be summarized by comparing an fBm of parameters H and S_0 to a WM function of parameters H, B, ν, κ_0. First of all, they both possess the same Hurst parameter, and hence, at least in the limit $\nu \to 1$, hold a self-affine behavior and the same fractal dimension. This is consistent with the fact that the equivalent power-spectral decay of the WM function and the power-spectral decay of the corresponding fBm process are the same. Finally, if B and κ_0 are selected according to Equation (3.56), then the power content of the WM function and the equivalent fBm process are equal on appropriate spectral intervals; in the limit of $\nu \to 1$, this last result is valid on any spectral interval.

The established link between WM functions and corresponding fBm processes generates a handy procedure to realize samples of band-limited fBm processes by using physical WM functions. It is not trivial to obtain realizations of fBm ensemble functions characterized by H and S_0 parameters. The alternative, simpler way consists of evaluating, via Equation (3.56), the B parameter for a corresponding WM function, whose ensemble elements are certainly easier to evaluate via Equation (3.47). In this equation, the H value is equal to the Hurst coefficient of the fBm process, selection of the ν value states how closely the WM discrete spectrum represents the corresponding fBm continuous one, and selection of κ_0 and P values are related to the process band limitation (see Equations [3.48] through [3.50]).

3.8. A Chosen Reference Fractal Surface for the Scattering Problem

In the following chapters, closed-form solutions to the scattering problem are derived. Any solution makes use of a fractal model to represent the natural-surfaces geometric properties, and allows evaluating the scattered electromagnetic field directly in terms of the surface fractal parameters and the illumination conditions. In those chapters, some comments are devoted to explaining the influence of each single fractal parameter on the scattered field; sometimes this is supported by inspection of the scattering diagrams. It is therefore convenient to introduce a reference surface to be employed to parametrically study the influence of each fractal parameter on the scattered field: this analysis is implemented by changing the fractal parameters with respect to those of the reference surface.

The selected fractal and illumination parameters identify a reference case somehow typical of the Earth environment as illuminated at

TABLE 3.2 Parameters relevant to the illumination conditions and to the reference surface considered in the following chapters to study the influence of the fractal parameters onto the scattered field.

Incidence angle (ϑ_i)	45°
Illuminated area (X, Y)	(1 m, 1 m)
Electromagnetic wavelength (λ)	0.1 cm
Hurst exponent (H)	0.8
Tone wave-number spacing coefficient (ν)	e
Overall amplitude-scaling factor (B)	0.01 m

microwave frequencies. In particular, all pertinent parameters selected for the reference case are reported in Table 3.2.

3.9. Fractal-Surface Models and their Comparison with Classical Ones

The comparison between classical and fractal models to represent two-dimensional surfaces is now presented. Some words of warning are in order.

Classical and fractal surfaces hold different topological dimensions. Then, by definition, a full correspondence is not achievable. There is no correspondence between classical surfaces and mathematical fractal ones. However, a link can be searched whenever a physical fractal is in order. In this case, the fractal surface is band-limited, and the fractal description may contain properties not belonging only to the surface.

Additional comments are due in order to discuss the complexity of classical and fractal description.

Classical surfaces are usually described by means of a regular process. Its description starts with the definition of a pdf for the surface height. If a link between heights for different values of the independent space variable is considered, then at least a second-order characterization is introduced. Hence, the classical approach would require specifying at least two functions (pdf and joint pdf). This description is classically very much simplified by postulating a zero-mean isotropic Gaussian pdf and a second-order Gaussian joint pdf: in this case, only a parameter σ and a function $R(\tau)$ are required. However, surfaces obtained by employing the classical approach do not resemble the natural-surfaces behavior. It can be argued that if these natural surfaces must be modeled, the classical approach has to be extended. Natural extension relies on using for the autocorrelation function more parameters or different shapes; alternatively, higher-order descriptions are employed. In any case, in the overall process description, other—in principle, infinite—parameters are introduced.

Fractal surfaces can be described by means of regular (fBm) as well as predictable (WM) processes. It has been shown (see Section 3.7) that the two processes can be linked. In the case that a regular process is employed, fractal-surfaces description starts with the definition of the pdf for the process of the height increments. At this stage, the process representing the height of the surface is not prescribed, and additional constraints must be imposed—for instance, a reference altitude (see Section 3.5.1). Natural surfaces have been shown to be well represented by means of this simple fractal description: for most applications, no other higher-order description is required.

Then a correspondence between classical and fractal models should first be searched, checking if both descriptions may prescribe the same σ value and the same autocorrelation or structure function. The value of σ for band-limited fractal models is evaluated in the next subsection, along with the standard deviation of the first and second derivative of the band-limited process. As far as the partial second-order description is concerned, it turns out that the shapes of the structure and autocorrelation functions for classical and fractal surfaces are different. The power-law behavior exhibited by the fractal model is not met by any classical model. As a matter of fact, the fractal model prescribes a power-law behavior at any scale, both in the spatial and in the spectral domain; and the fractal parameters allow us to select the shapes of these power-law behaviors. Conversely, the power-law autocorrelation introduced in Section 2.A.4 actually provides a power-law behavior in the space domain only for large distances; similarly, the exponential autocorrelation in Section 2.A.2 actually provides a power-law behavior in the spectral domain only for large wavenumbers. Moreover, the shapes of these power-law behaviors are fixed and cannot be selected by an appropriate parameter. In conclusion, to assure model flexibility, the classical models may host even a large number of parameters, but they never play the same role played by the fractal ones.

It turns out that even for band-limited fractal surfaces, a full correspondence is not achievable, because, in general, the number of independent parameters is not the same for the classical-surface candidate to be put into

3.9. Fractal-Surface Models and their Comparison with Classical Ones

correspondence with the fractal one. This limitation is crucial. A fractal description relies on few independent parameters: two parameters are needed for mathematical fBm, whereas two additional parameters are required to band-limit the process.

Simple classical functions are also described in terms of few parameters, two in the case of Gaussian or exponential distributions; but these descriptions very poorly resemble natural surfaces. An improved correspondence may be obtained, at the expense of introduction of additional (in principle, very large) number of parameters, thus making a comparison with the fractal description very difficult.

Summarizing, it can be stated that the classical description of the geometric properties of surfaces requires, in general, either a multiparameter or a multifunctional approach to render the model accurate. On the contrary, the fractal description relies on a limited number of parameters, but implies dealing with a number of issues (e.g., the infrared catastrophe) that strongly complicate its use to analytically obtain closed-form solution to the scattered field. These shortcoming may be mitigated if physical fractals are considered, which requires introduction of additional parameters (bandwidth limits), but may render the description dependent also on the sensor characteristics.

3.9.1. Classical Parameters for the fBm Process

At variance to mathematical, not stationary, fBm surfaces, for physical band-limited stationary fBm fractal surfaces, some classical-surface parameters can be evaluated within the limitations stated at the beginning of this section. Considering the surface-height stochastic process and its first and second derivative, the corresponding variances, σ^2, σ'^2 and σ''^2 can be computed; their values can be obtained directly from the process power spectrum $W(\kappa_x, \kappa_y)$ as follows:

$$\sigma^2 = \frac{1}{4\pi^2} \iint W(\kappa_x, \kappa_y) d\kappa_x d\kappa_y, \quad (3.58)$$

$$\sigma'^2 = \frac{1}{4\pi^2} \iint \left(\kappa_x^2 + \kappa_y^2\right) W(\kappa_x, \kappa_y) d\kappa_x d\kappa_y, \quad (3.59)$$

$$\sigma''^2 = \frac{1}{4\pi^2} \iint \left(\kappa_x^2 + \kappa_y^2\right)^2 W(\kappa_x, \kappa_y) d\kappa_x d\kappa_y. \quad (3.60)$$

The spectral-power density to be included in Equations (3.58) through (3.60) is relevant to the band-limited fractal surface, and is provided

TABLE 3.3 Variances of surface height (σ^2), slope (σ'^2), and curvature (K^2) for a band-limited fBm surface, as a function of S_0, α (2nd column) or T, H (3rd column).

Variances	Expressions	
σ^2	$\dfrac{S_0}{2\pi(\alpha-2)}\left(\dfrac{1}{\kappa_m^{\alpha-2}}-\dfrac{1}{\kappa_M^{\alpha-2}}\right)$	$2^{2H-1}\Gamma^2(1+H)\dfrac{\sin(\pi H)}{\pi H}T^2$ $\left[\left(\dfrac{\tau_M}{T}\right)^{2H}-\left(\dfrac{\tau_m}{T}\right)^{2H}\right]$
σ'^2	$\dfrac{S_0\left(\kappa_M^{4-\alpha}-\kappa_m^{4-\alpha}\right)}{2\pi(4-\alpha)}$	$2^{2H-1}\Gamma^2(1+H)\dfrac{\sin(\pi H)}{\pi(1-H)}$ $\left[\left(\dfrac{T}{\tau_m}\right)^{2-2H}-\left(\dfrac{T}{\tau_M}\right)^{2-2H}\right]$
$\sigma''^2 \cong K^2$	$\dfrac{S_0\left(\kappa_M^{6-\alpha}-\kappa_m^{6-\alpha}\right)}{2\pi(6-\alpha)}$	$2^{2H-1}\Gamma^2(1+H)\dfrac{\sin(\pi H)}{\pi(2-H)}\dfrac{1}{T^2}$ $\left[\left(\dfrac{T}{\tau_m}\right)^{4-2H}-\left(\dfrac{T}{\tau_M}\right)^{4-2H}\right]$

by Equation (3.40): the power-spectral density is given by Equations (3.25) through (3.27) in the range $\kappa_m < \kappa < \kappa_M$; outside this interval, the power spectrum is taken equal to zero. In Table 3.3, evaluations of these parameters in terms of the spectral-domain (amplitude and slope) and spatial-domain parameters (topothesy and Hurst coefficient) are reported.

Evaluations of σ^2, σ'^2 and σ''^2 are relevant because those classical parameters are indicators of the surface behavior without formally requiring knowledge of the autocorrelation function: this implies that they can also be defined if the surface has autocorrelation that is not only a function of the space lag. Moreover, σ^2, σ'^2, and σ''^2 also play a fundamental role in evaluating the electromagnetic scattering from classical surfaces: as a matter of fact, σ^2 is the surface variance, and usually enters directly into the scattered-field expression; σ'^2 represents the surface slope, and is normally used to assess the scattering-solution convergence; σ''^2 provides an approximation for the surface curvature K^2, assuming that the surface slope is small, and enters into the approximation to be selected for the scattered-field evaluation. Note that the radius of curvature is defined as $1/K$.

Many direct in situ measurements over natural—soil and ocean—surfaces verified that the latter are self-affine, in agreement with Equation (3.9); that they exhibit power-law spectra, in agreement with Equation (3.25); and that the measured height standard deviation is proportional to the length of the considered profile raised to a power greater than 1 (for $H > 1/2$, see Section 3.5.1), in agreement with the evaluation of the surface variance reported in Table 3.3. These results validate the use of fBm for modeling natural surfaces.

3.9.2. Classical Parameters for the WM Function

It is assumed that the C_ps are mutually independent, zero-mean, unitary-variance Gaussian random variables; and that the Ψ_ps as well as the Φ_ps are mutually independent, uniformly distributed in $[-\pi, \pi)$ random variables. It turns out from Equation (3.47) that the surface mean is zero, and the surface variance σ^2 is equal to

$$\sigma^2 = \langle z^2(x,y) \rangle = \frac{1}{2}B^2 \sum_{p=0}^{P-1} \nu^{-2Hp} = \frac{B^2(1-\nu^{-2HP})}{2(1-\nu^{-2H})}. \quad (3.61)$$

Note that setting of the B value allows generating a surface with any prescribed variance.

For the other variances employed in evaluating the electromagnetic scattering from classical surfaces, it turns out that

$$\sigma'^2 = \langle |\nabla z|^2 \rangle, \quad (3.62)$$

$$\sigma''^2 = \langle (\nabla^2 z)^2 \rangle. \quad (3.63)$$

All results are collected in Table 3.4. Evaluation of σ''^2 provides a good approximation for the curvature K^2 if surface slope is small.

Finally, it is worth noting that in the limit $\nu \to 1$, variances of surface height, slope, and curvature for a band-limited WM surface, in Table 3.4, are in agreement with the corresponding parameters for a band-limited fBm surface in Table 3.3, as can be verified by using results of Sections 3.6 and 3.7.

TABLE 3.4 Variances of surface height (σ^2), slope (σ'^2), and curvature (K^2) for a band-limited WM surface.

Variances	Expressions
σ^2	$\dfrac{B^2(1-\nu^{-2HP})}{2(1-\nu^{-2H})}$
σ'^2	$\kappa_0^2 \dfrac{B^2}{4} \dfrac{1-\nu^{-2(H-1)P}}{1-\nu^{-2(H-1)}}$
$\sigma''^2 \cong K^2$	$\kappa_0^4 \dfrac{B^2}{8} \dfrac{1-\nu^{-2(H-2)P}}{1-\nu^{-2(H-2)}}$

3.10. References and Further Readings

Mandelbrot's book (1983) introduces fractal geometry: philosophy beyond fractal geometry is deeply illustrated, and the mathematical implications are dealt with in detail; attention is also focused on mathematical fractals. The work of Falconer (1990) presents fractals in all mathematical aspects, and the reader can assimilate some fundamental mathematical instruments required to deal with fractal geometry. Several chapters are devoted to illustrate mathematical properties of fractal surfaces. Surfaces of fBm type are presented in a more intuitive way in a study by Feder (1988). In particular, the books by Mandelbrot and Falconer are starting references to assess the fundamental question of existence of stochastic processes holding the fBm properties. In these same works, proofs of the relationship between fractal dimension and the Hurst coefficient are reported. In situ measurements over natural soil and ocean surfaces, verifying that the latter are modeled by means of fBm processes, can be found in Mandelbrot's work. An in-depth presentation of the WM function can be found in a study by Berry and Lewis (1980), where the WM fractal dimension is discussed.

Appendix 3.A Generalized Functions

Generalized functions are the generalization of the classical concept of mathematical function. Generalized functions are alternatively called *distributions*, because from the physical viewpoint they are introduced and are used to ideally describe the distributions of some physical quantities.

Hence, use of generalized functions allows us to express in a mathematically correct form idealized physical concepts such as the density of a material point, the point charge or source, the space-charge density of a layer, the intensity of an instantaneous charge or source, and so on. A very popular distribution is the Dirac δ-function: for instance, the charge in a volume $q(\mathbf{r})$ is represented as superposition of point charges at points $\mathbf{r}'$:

$$q(\mathbf{r}) = \int_{-\infty}^{\infty} q(\mathbf{r}') \delta(\mathbf{r} - \mathbf{r}') d\mathbf{r}' \qquad (3.A.1)$$

In general, a distribution f is a continuous linear function defined on an appropriate space of test function φ, $f: \varphi \to (f, \varphi)$.

Definition of the distributions on Ω, an open set of R^k, is in order. A key space for the test functions is $D(\Omega)$, represented by functions with compact support, differentiable at any order and assuming values on the complex plane C (i.e., $\varphi: \Omega \to C$). Then, a distribution on Ω is a continuous linear function $f: D(\Omega) \to C$; the space for the distribution is $D'(\Omega)$, dual space to $D(\Omega)$.

For a generalized function, the FT is defined. In this case, the space for the test functions requires the existence of the FT of φ

$$\text{FT}[\varphi] = \int_{-\infty}^{\infty} \varphi(\mathbf{r}) \exp(-i\mathbf{k} \cdot \mathbf{r}) d\mathbf{r} \qquad (3.A.2)$$

and the FT of f is defined as

$$(\text{FT}[f], \varphi) \stackrel{\Delta}{=} (f, \text{FT}[\varphi]). \qquad (3.A.3)$$

Hence, the Fourier transform of a nonintegrable function is allowed in the sense of distributions, providing that the test functions belong to a convenient set.

Appendix 3.B Space-Frequency and Space-Scale Analysis of Nonstationary Signals

3.B.1. Introduction

Application of the fractal models often requires estimating the fractal parameters of a prescribed surface. Then, appropriate inverse techniques are designed. Independently from the selected technique, the assumptions given under Section 3.5 are no longer valid: the surface under analysis is generally viewed only on a finite extent, or its acquisition is provided only up to fixed scales. Then, the measurement operations lead to an estimated spectrum that differs from the mathematical fractal one, and it is necessary to operate in such a way as to avoid the possibility that the estimation procedure could lead to wrong fractal parameters.

Obtaining a spectral representation for an fBm surface is a delicate point—because the fBm is not stationary, and suffers from the infrared catastrophe. Techniques to handle nonstationary signals are referred to as *time-frequency* and *time-scale* analysis, respectively, whenever one-dimensional signals are in order and the independent variable is time. The reader can make reference to the literature in the field to find complete mathematical treatments along with relevant proofs of the material reported below.

Stationary signals can be studied by making reference to relative distances only of the independent variables. The simplest manipulations that can be performed consist of applying to the signal under analysis appropriate linear and bilinear transforms. In the first case, signal decomposition is in order, and a popular approach consists of Fourier transforming the signal with respect to the independent variables; in the second case, energy distribution is in order, and a popular approach consists of evaluating the signal power-density spectrum (PDS); the two approaches are obviously related. Stationary signals lead to FT and PDS that do not depend on the independent variables.

It is almost intuitive that whenever nonstationary signals are considered, the independent variables must enter with their values, not only with their relative differences. In this case, linear and bilinear transforms map the nonstationary signal whose independent variable is, for instance, space into a new function that depends on the same space variable present in the original signal and on a new auxiliary (possibly vector) variable. In the *space-wavenumber* representations, the auxiliary variable is a wave vector; in the *space-scale* representations, the auxiliary variable is a scale factor.

3.B.2. fBm Wigner-Ville Spectrum

The power spectrum of a nonstationary process can be defined following a space-wavenumber approach via the *Wigner-Ville spectrum*: this is done in the following discussion for the fBm.

Appendix 3.B Space-Frequency and Space-Scale Analysis

Equations (3.17) and (3.18) can be formally restated using the Wigner-Ville spectrum $W(\kappa, \mathbf{r})$, defined for deterministic signals as

$$W(\mathbf{r}, \kappa) \stackrel{\Delta}{=} \int_{-\infty}^{+\infty} z\left(\mathbf{r} + \frac{\tau}{2}\right) z^*\left(\mathbf{r} - \frac{\tau}{2}\right) \exp[-i\kappa \cdot \tau] d\tau; \qquad (3.B.1)$$

applying the stochastic average to Equation (3.B.1) and exchanging the integration limits, it turns out that for the topological two-dimensional case, which is of our interest:

$$W(\kappa) = \lim_{q \to \infty} \left(\frac{1}{2q}\right)^2 \int_{-q}^{q} d\mathbf{r} W(\kappa, \mathbf{r}). \qquad (3.B.2)$$

Whenever process realizations of finite extent are available, and if these include details of the surface at any scale, then only the Wigner-Ville spectrum can be evaluated. In this case, the power-density spectrum is obtained only via a limit operation whenever the area under analysis is taken to be of infinite extent.

Equations (3.23) and (3.B.2) also allows evaluating the Wigner-Ville spectrum of the fBm in closed form:

$$W(\kappa, \mathbf{r}) = \frac{T^{2(1-H)}}{2} 2^{2+2H} \Gamma^2 (1+H) \sin(\pi H) \frac{1}{\kappa^{2+2H}}$$

$$\left[1 - \frac{1}{2^{2H-1}} \cos(2\kappa \cdot \mathbf{r})\right]. \qquad (3.B.3)$$

Results in Equation (3.B.3) helps interpreting Equations (3.24) and (3.25) relevant to the power density spectrum of an fBm surface. Then, for a stochastic process of finite extent, the Wigner-Ville spectrum oscillates with respect to the space variable around an overall power-law behavior. This behavior provides a rationale to interpret fBm spectra obtained from measurements over surface samples of finite extent.

Equation (3.25) is amenable to both a mathematical and a physical interpretation. From the mathematical viewpoint, the power-density spectrum of an fBm process holds a power-law behavior; moreover, the infrared catastrophe is attained because the power-density spectrum is not integrable in the low-frequency range. From the physical viewpoint, Equation (3.24) shows that the power-law behavior is asymptotically approximated for $q \to \infty$: because q limits the integration interval over the space variable, $q \to \infty$ means that the asymptotic behavior is attained for long space intervals.

3.B.3. fBm and Wavelet Approach

The power spectrum of a nonstationary process can be defined following a space-scale approach via the *wavelet transform*: this is done in the following section for the fBm.

A spatial-filter function $g(\cdot)$ is applied to the fBm process: the filtered surface z^{filt} is represented as follows:

$$z^{filt}(\mathbf{r}, a) = \frac{1}{a} \int_{-\infty}^{+\infty} z(\mathbf{t}) \, g\left(\frac{\mathbf{t}-\mathbf{r}}{a}\right) d\mathbf{t}, \quad (3.B.4)$$

wherein $a > 0$, and $g(\mathbf{r})$ is the inverse FT of the filter function $G(\kappa)$. Integration is extended to the 2-D real plane. Equation (3.B.4) is the *wavelet transform* of the unfiltered surface $z(\cdot)$: necessary and sufficient conditions for this transformation to be invertible are provided by the *admissibility condition*.

The filtered surface has zero mean, as follows by computing the statistical average of Equation (3.B.4): statistical average and integral operation are exchanged, and $\langle z^{filt}(\mathbf{t}) \rangle = 0$ because also the unfiltered fBm has zero mean upon setting $z(0) = 0$.

Evaluation of the filtered-surface autocorrelation is now in order:

$$R^{filt}(\mathbf{r}, \mathbf{r}', a) = \langle z^{filt}(\mathbf{r}, a) \, z^{filt}(\mathbf{r}', a) \rangle$$

$$= \frac{T^{2(1-H)}}{2} \int_{-\infty}^{+\infty} \int_{-\infty}^{+\infty} \left(|\mathbf{t}|^{2H} + |\mathbf{t}'|^{2H} - |\mathbf{t}-\mathbf{t}'|^{2H} \right)$$

$$g\left(\frac{\mathbf{t}-\mathbf{r}}{a}\right) g\left(\frac{\mathbf{t}'-\mathbf{r}'}{a}\right) d\mathbf{t} \, d\mathbf{t}', \quad (3.B.5)$$

where Equation (3.B.4) has been used, as well as statistical-average implementation and integrals exchange, and Equation (3.11) accounted for.

Assume now that $g(\mathbf{t}/a)$ is a real even function of $\mathbf{t}$, and is either localized—that is, has a compact support—or infinitesimal of order greater than $2H + 2$ for $\mathbf{t} \to \infty$. Then it can be shown that the integral of the first two factors in Equation (3.B.5) is rendered equal to zero. Accordingly, the process becomes stationary, and the autocorrelation function attains the expression

Appendix 3.B Space-Frequency and Space-Scale Analysis

$$R^{filt}(\mathbf{r},\mathbf{r}',a) = \frac{T^{2(1-H)}}{2} a^{2H+2} \int_{-\infty}^{+\infty} |\xi|^{2H} d\xi$$

$$\int_{-\infty}^{+\infty} g\left(\frac{t'-\mathbf{r}'}{a}\right) g\left(\frac{t'-a\xi-\mathbf{r}}{a}\right) dt'$$

$$= \frac{T^{2(1-H)}}{2} a^{2H+2} \int_{-\infty}^{+\infty} |\xi|^{2H} d\xi \int_{-\infty}^{+\infty} g(\eta)$$

$$g\left(\eta - \xi + \frac{\mathbf{r}'-\mathbf{r}}{a}\right) d\eta, \qquad (3.B.6)$$

where the coordinates transformation $\xi = (t'-t)/a$, $\eta = (t'-\mathbf{r}')/a$ has been implemented. Equation (3.B.6) can be rewritten as

$$R^{filt}(\mathbf{r}-\mathbf{r}',a) = \frac{T^{2(1-H)}}{2} a^{2H+2} \int_{-\infty}^{+\infty} |\xi|^{2H} R_g\left(\xi - \frac{\mathbf{r}'-\mathbf{r}}{a}\right) d\xi, \qquad (3.B.7)$$

wherein the even property of the autocorrelation function with respect to the space variable has been accounted for and

$$R_g(\xi) = \int_{-\infty}^{+\infty} g(\eta) g(\eta - \xi) d\eta \qquad (3.B.8)$$

is the autocorrelation of the filter function $g(\cdot)$.

The FT of Equation (3.B.7) with respect to $\mathbf{r}' - \mathbf{r}$ provides the fBm power spectrum filtered at the scale a. The FT of $R_g(\xi)$, which is itself a convolution, is just $G^2(\kappa)$, once $g(\cdot)$ is selected real and even; the FT of the other factor, $|\xi|^{2H}$, is provided in Equation (3.21). Then the power spectrum of the filtered surface is

$$W^{filt}(\kappa,a) = \frac{T^{2(1-H)}}{2} a^{2H+2} aG^2(ak) \left\{ \text{FT}\left[|\xi|^{2H}\right]\Big|_{ak} \right\}$$

$$= \frac{T^{2(1-H)}}{2} a^{2H+2} aG^2(ak) 2^{2+2H} \Gamma^2(1+H) \sin(\pi H) \frac{1}{(ak)^{2+2H}}$$

$$= aG^2(ak) \frac{S_0}{k^{2+2H}}, \qquad (3.B.9)$$

wherein the scaling theorem for the Fourier transform, $\text{FT}[af(t)] = |a|^{-1} F(a\omega)$, has been accounted for, because the convolution in Equation (3.B.7) is evaluated in $\eta = (\mathbf{r}'-\mathbf{r})/a$, and S_0 has been provided by Equation (3.26).

The power spectrum of the fBm surface, $W(\kappa)$, can be finally recovered by applying the energy theorem for the wavelet transform:

$$W(\kappa) = \int_0^{+\infty} W^{filt}(\kappa, a) \frac{1}{a^2} da; \qquad (3.B.10)$$

then, considering the admissibility condition on the WT, it turns out that

$$W(\kappa) = \frac{S_0}{\kappa^{2+2H}}, \qquad (3.B.11)$$

thus confirming the results previously obtained with a different approach and reported in Equations (3.25) and (3.27).

Examination of Equation (3.B.9) allows us to identify its last factor as the power spectrum of the unfiltered surface, as anticipated in Section 3.5.1. However, its generalized FT, $|\mathbf{r}|^{2H}$, is not the autocorrelation function of the unfiltered surface, because this latter is not stationary (see Equation [3.11]).

The filtered spectrum again has a clear mathematical and physical counterpart for nonstationary signals: it is related to a convenient linear transform that can be applied to stochastic processes, and provides an energy distribution of the stochastic process that is related both to space and scale. Obviously, it is also related to the implemented filter-transfer function. Whenever process realizations of infinite extent in a finite number of scales are in order, only the filtered spectrum can be evaluated, and the power-density spectrum is obtained only via a further integration over all the scales. If the filter function holds some relevant energy properties that are conveniently formalized within the wavelet approach, the obtained fBm spectrum does not depend on the filter-transfer function, and represents only the fBm behavior.

3.B.3.1. Filter Implementation

A surface is observed, sensed, or measured to determine its parameters according to a postulate model (*analysis procedure*). It is explored over a finite extent with a sensor of finite resolution: the filter is automatically implemented, and the analyzed surface turns out to be stationary. The dual case (*synthesis procedure*) consists of the assumption of the parameters of the chosen model to generate, or study, realizations of the surface ensemble. Then the filter procedure must be implemented if a stationary surface is desired.

In the analysis procedures, the filter-transfer function, $G(\kappa)$, is prescribed by the selected observing, sensing, or measuring instrument. If the filtered

Appendix 3.B Space-Frequency and Space-Scale Analysis

and unfiltered surface should hold the same power-density spectrum in the observation bandwidth $[\kappa_m, \kappa_M]$, additional constraint must be enforced to the filter function. Admissibility conditions should be satisfied to assure that the filter does not corrupt the fBm spectrum; however, if the overall spectral amplitude is not of interest, simpler choices are allowed. In the case of isotropic surfaces, a wide class of admissible, commonly encountered filters can attain the form

$$\begin{cases} G(0) = 0 \\ |G(\kappa)| < 1 & \forall \kappa: \kappa < \kappa_m \\ |G(\kappa)| = 1 & \forall \kappa: \kappa_m < \kappa < \kappa_M, \\ |G(\kappa)| < 1 & \forall \kappa: \kappa > \kappa_M \\ G(\kappa) \to 0 & \forall \kappa: \kappa \to \infty \end{cases} \quad (3.B.12)$$

being everywhere continuous and sufficiently smooth in the origin to assure that the filter-response function $g(\mathbf{r})$ has compact support or is infinitesimal of order greater than $2H+2$ for $|\mathbf{r}| \to \infty$. This filter function does not alter the spectral behavior for the wave-number band of interest $\kappa_m < \kappa < \kappa_M$. Moreover, it is convenient to select a filter-transfer function $G(\kappa)$ such that the integral of $W^{filt}(\kappa)$ in the transition regions, $[0, \kappa_m]$ and $[\kappa_M, \infty]$, is much lower than that in the observation bandwidth $[\kappa_m, \kappa_M]$.

In the synthesis procedure, the shape of the filter function can be selected to be as simple as possible: $a = 1$ is first set in the Equations included in Section 3.B.3, the simple filter is implemented via Equation (3.34) and, by recalling the result of Section 3.5.2, the filtered fBm turns out to be stationary.

CHAPTER 4

Analytic Formulations of Electromagnetic Scattering

4.1. Introduction and Chapter Outline

Exact solution to the problem of electromagnetic-wave scattering must rely on Maxwell equations together with the associate boundary conditions. In this book, reference is made to the phasor domain, referring to time-harmonic electromagnetic fields. Corresponding time-domain fields are recovered taking the real part of the phasor times $\exp[i\omega t]$, where $i = \sqrt{-1}$, $\omega = 2\pi f$, f being the electromagnetic-wave frequency, and t the time. In passim, the other convention $\exp[-i\omega t]$ is also followed in the current literature: the latter is obtained from the adopted one by changing the sign to the imaginary unit i. This is not always an automatic procedure: for instance, Hankel functions of second kind must be replaced by Hankel functions of first kind.

In general, no analytic closed-form exact solution to the scattered field is achievable for natural surfaces. Only approximate analytic solutions are obtained by implementing some simple models for the boundary conditions relevant to the scattering surface; different approximations lead to different methods to evaluate the scattered field, each method holding under the appropriate surface roughness regime. Only for very simple natural-surface models do these approximate methods lead to an analytic closed-form solution for the scattered field.

116   4 ◊ Analytic Formulations of Electromagnetic Scattering

In this chapter, the rationale for expressing the scattered field in analytic form is presented. Starting from Maxwell equations (Section 4.2), the scattering problem is formulated according to the integral-equation method (Section 4.3), whose implementation requires a certain number of steps. The first step is the choice of a reference system that conveniently allows dealing with field polarization (Section 4.4); then, two frameworks are presented where closed-form solutions to the scattering problem can be gained. One of the frameworks is referred to as the *Kirchhoff Approximation* (KA) (Section 4.5), and leads to the *Physical-Optics* (PO) solution (Section 4.6). The other framework is the *Extended-Boundary-Condition Method* (EBCM) (Section 4.7), which leads to the *Small-Perturbation Method* (SPM) (Section 4.8).

The general theoretical background is presented before any choice on the evaluation techniques and the surface models is detailed. Hence, the reported material is general, can be applied to any surface, and is preliminary to the following chapters from 5 to 8: in Chapters 5 and 6, the analytic solution is found within the KA, whereas Chapters 7 and 8 refer to EBCM. In Chapters 5 and 7, the WM process is used to model the geometric surface; conversely, in Chapters 6 and 8, the adopted surface geometric model is the fBm process. Applications of presented methods to the classical surfaces are found in several excellent books, and reference is made to them for a full discussion.

4.2. Maxwell Equations

Electromagnetic fields are postulated to satisfy Maxwell equations. In time domain, their differential form reads as

$$\begin{cases} \nabla \times \mathbf{e} = -\dfrac{\partial \mathbf{b}}{\partial t} \\ \nabla \times \mathbf{h} = \dfrac{\partial \mathbf{d}}{\partial t} + \mathbf{j}, \\ \nabla \cdot \mathbf{d} = \rho \\ \nabla \cdot \mathbf{b} = 0 \end{cases} \quad (4.1)$$

wherein t [s] is the time variable, $\mathbf{e}$ [V/m] and $\mathbf{h}$ [A/m] are the electric and magnetic fields, $\mathbf{b}$ [Wb/m^2] and $\mathbf{d}$ [F/m^2] the electric and magnetic inductions, and $\mathbf{j}$ [A/m^2] and ρ [C/m^3] the electric current and charges densities, respectively. All these functions are space-time-dependent fields.

4.2. Maxwell Equations

In the Fourier domain, Equations (4.1) specify as follows:

$$\begin{cases} \nabla \times \mathbf{E} = -i\omega \mathbf{B} \\ \nabla \times \mathbf{H} = i\omega \mathbf{D} + \mathbf{J} \\ \nabla \cdot \mathbf{D} = P \\ \nabla \cdot \mathbf{B} = 0 \end{cases}, \quad (4.2)$$

wherein $\omega = 2\pi f$, f [s^{-1}] being the frequency, and each field in the transformed domain is indicated with the corresponding capital letter; their dimensions are those in time domain multiplied by the dimension time. All these functions are ω-**r**-dependent fields. In the phasor domain, the sinusoidal dependence of the fields is prescribed: the same Equations (4.2) are obtained, ω is a constant, fields depend on the position only, and their dimensions are equal to those of the corresponding observables in time domain.

Equations (4.2) do not include any dependence on the medium properties: those are specified by the constitutive relations that relate inductions and currents to the fields. For linear, time-invariant, spatially nondispersive media, the constitutive relations usually relate each induction only to the corresponding field. Hence, in the phasor domain, they can be expressed as follows:

$$\begin{cases} \mathbf{D}(\mathbf{r}, \omega) = \boldsymbol{\varepsilon}(\mathbf{r}, \omega) \cdot \mathbf{E}(\mathbf{r}, \omega) \\ \mathbf{B}(\mathbf{r}, \omega) = \boldsymbol{\mu}(\mathbf{r}, \omega) \cdot \mathbf{H}(\mathbf{r}, \omega), \\ \mathbf{J}(\mathbf{r}, \omega) = \boldsymbol{\sigma}(\mathbf{r}, \omega) \cdot \mathbf{E}(\mathbf{r}, \omega) \end{cases} \quad (4.3)$$

wherein $\boldsymbol{\varepsilon}$, $\boldsymbol{\mu}$ and $\boldsymbol{\sigma}$ are the permittivity, permeability, and conductivity of the medium, respectively. In general, these are represented by 3×3 matrices, and reduce to scalars whenever isotropy is invoked. In this case, Maxwell equations take the form

$$\begin{cases} \nabla \times \mathbf{E} = -i\omega\mu \mathbf{H} \\ \nabla \times \mathbf{H} = i\omega\varepsilon \mathbf{E} + \mathbf{J} \\ \nabla \cdot \varepsilon\mathbf{E} = P \\ \nabla \cdot \mu\mathbf{H} = 0 \end{cases}, \quad (4.4)$$

wherein the conductivity is usually included within a complex permittivity:

$$\varepsilon \to \varepsilon - i\frac{\sigma}{\omega}, \quad (4.5)$$

so that **J** and *P* stand only for current and charge sources. Ohmic, dielectric, and magnetic losses are accounted for by the imaginary part of permeability and permittivity.

4.3. The Integral-Equation Method

Solution of Maxwell equations implies managing a three-dimensional integral problem: all unknowns are defined in a volume. Whenever the space under analysis can be decomposed in homogenous regions, this may lead to two-dimensional integral equations whose unknowns must be evaluated only on two-dimensional surfaces. Hence, one advantage of the surface integral-equation method relies on reducing by a factor of 1 the problem dimensionality, because the unknowns appear only in surface integrals.

The integral-equation method is summarized as follows. For each homogeneous region, the Green's function is known. Then, in each region, the field is expressed as the superposition of two contributions: radiation by the (volume) sources in the region, and by equivalent surface sources at the region interfaces. These surface sources are the problem unknowns: they are determined by solving the surface integral equations that are obtained by applying boundary conditions at the interfaces between different homogeneous regions. The procedure is quite general, and applies to a wide set of problems: scalar as well as vector fields can be dealt with; moreover, even for nonhomogeneous regions, surface integral equations can be derived.

In the following paragraphs, the surface integral equations are set for the problem of interest (see Figure 4.1): the surface S separates two regions V_1 and V_2, each one filled with homogeneous media; the sources are confined in the region V_1. The field scattered in the region V_1 (and transmitted in the region V_2) must be computed.

For each homogeneous region, the wave equation for the electric field, in phasor domain, is obtained by taking the curl of the first of Maxwell Equations (4.4) and substituting the second one:

$$\nabla \times \nabla \times \mathbf{E}_1(\mathbf{r}) - \omega^2 \varepsilon_1 \mu_1 \mathbf{E}_1(\mathbf{r}) = -i\omega \mu_1 \mathbf{J}(\mathbf{r}), \ \mathbf{r} \in V_1 \quad (4.6)$$

$$\nabla \times \nabla \times \mathbf{E}_2(\mathbf{r}) - \omega^2 \varepsilon_2 \mu_2 \mathbf{E}_2(\mathbf{r}) = 0, \ \mathbf{r} \in V_2, \quad (4.7)$$

wherein, $\mathbf{E}_{1,2}, \varepsilon_{1,2} \mu_{1,2}$, are the electric fields, permittivity, and permeability in the two media, respectively. For the sake of simplicity, here and in the

4.3. The Integral-Equation Method

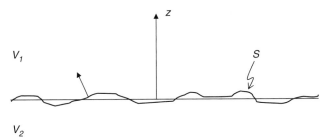

FIGURE 4.1 Geometry of the scattering problem projected onto a plane containing the z-axis. The scattering surface S separating the two homogeneous regions V_1 and V_2 is shown. The surface outgoing normal is depicted at a generic point of the surface.

following paragraphs, isotropic media are considered, so that permittivity and permeability are represented by scalar quantities.

To proceed further, radiation from elementary sources is in order. In the frequency domain, the field radiated at $\mathbf{r}$ by an elementary source $\mathbf{s}$ set at $\mathbf{r}'$ is given by $\mathbf{G}(\mathbf{r}-\mathbf{r}') \cdot \mathbf{s}(\mathbf{r}')$, wherein $\mathbf{G}(\mathbf{r}-\mathbf{r}')$ is the *dyadic Green's function*. For the two homogeneous media, the Green's functions are defined by

$$\nabla \times \nabla \times \mathbf{G}_1(\mathbf{r}-\mathbf{r}') - \omega^2 \varepsilon_1 \mu_1 \mathbf{G}_1(\mathbf{r}-\mathbf{r}') = \mathbf{I}\delta(\mathbf{r}-\mathbf{r}'), \quad (4.8)$$

$$\nabla \times \nabla \times \mathbf{G}_2(\mathbf{r}-\mathbf{r}') - \omega^2 \varepsilon_2 \mu_2 \mathbf{G}_2(\mathbf{r}-\mathbf{r}') = \mathbf{I}\delta(\mathbf{r}-\mathbf{r}'), \quad (4.9)$$

wherein $\mathbf{I}$ is the unitary dyadic.

Solution to Equations (4.8) and (4.9) is known and of the type

$$\mathbf{G}(\mathbf{r}-\mathbf{r}') = \left[\mathbf{I} + \frac{\nabla\nabla}{k^2}\right] \frac{\exp(-ik|\mathbf{r}-\mathbf{r}'|)}{4\pi|\mathbf{r}-\mathbf{r}'|}, \quad (4.10)$$

wherein $k = \omega\sqrt{\varepsilon\mu}$ is the wave number of the homogeneous isotropic space under consideration, and radiation condition at infinity has been enforced.

Due to reciprocity, the dyadic Green's function holds the following relevant properties, which also can easily be verified by inspection of Equation (4.10):

$$\mathbf{G}^t(\mathbf{r}'-\mathbf{r}) = \mathbf{G}(\mathbf{r}-\mathbf{r}'), \quad (4.11)$$

$$\left[\nabla \times \mathbf{G}(\mathbf{r}'-\mathbf{r})\right]^t = -\nabla \times \mathbf{G}(\mathbf{r}-\mathbf{r}'), \quad (4.12)$$

the suffix t indicating the transpose matrix.

Equations (4.6) through (4.9) must be appropriately combined and integrated to obtain four relevant integral equations. This procedure is illustrated in the following discussion.

In Equations (4.6) and (4.8), the variables $\mathbf{r}$ and $\mathbf{r}'$ are first exchanged with each other. Then the new Equation (4.6) is postmultiplied by $\mathbf{G}_1(\mathbf{r}' - \mathbf{r})$, whereas the new Equation (4.8) is premultiplied by $\mathbf{E}(\mathbf{r}')$; the difference between these two expressions is integrated over the volume V_1 (see Figure 4.2). The final result, valid $\forall \mathbf{r} \in V_1$, is the following one:

$$\int_{V_1} \left[\nabla' \times \nabla' \times \mathbf{E}_1(\mathbf{r}') \cdot \mathbf{G}_1(\mathbf{r}' - \mathbf{r}) - \mathbf{E}_1(\mathbf{r}') \cdot \nabla' \times \nabla' \times \mathbf{G}_1(\mathbf{r}' - \mathbf{r}) \right] dV'$$

$$= -i\omega\mu_1 \int_{V_1} \mathbf{J}(\mathbf{r}') \cdot \mathbf{G}_1(\mathbf{r}' - \mathbf{r}) dV' - \mathbf{E}_1(\mathbf{r}), \qquad (4.13)$$

where the factor containing the integral at the right side of the equation is recognized to be the incident field $\mathbf{E}^{(i)}(\mathbf{r})$ radiated by the sources in the volume V_1:

$$\mathbf{E}^{(i)}(\mathbf{r}) = -i\omega\mu_1 \int_{V_1} \mathbf{J}(\mathbf{r}') \cdot \mathbf{G}_1(\mathbf{r}' - \mathbf{r}) \, dV'$$

$$= -i\omega\mu_1 \int_{V_1} \mathbf{G}_1(\mathbf{r} - \mathbf{r}') \cdot \mathbf{J}(\mathbf{r}') dV', \qquad (4.14)$$

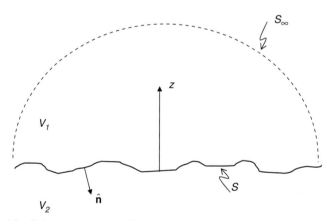

FIGURE 4.2 Relevant to the application of the dyadic divergence theorem to the volume V_1. The region V_1 is closed by means of the surfaces S and S_∞. The adopted normal is depicted at a generic point of the surface S.

4.3. The Integral-Equation Method

where Equation (4.11) has been accounted for. Note that in Equation (4.14) we switch from row vectors, as in Equation (4.13), to column vectors.

The volume integral in Equation (4.13) can be transformed onto a surface integral (see Figure 4.2) by using the dyadic Green's theorem, Equation (A.4.1) integrated over V_1, with $\mathbf{A} = \mathbf{E}_1$ and $\mathbf{B} = \mathbf{G}_1$:

$$\int_{V_1} \left[\nabla' \times \nabla' \times \mathbf{E}_1(\mathbf{r}) \cdot \mathbf{G}_1(\mathbf{r}' - \mathbf{r}) - \mathbf{E}_1(\mathbf{r}') \cdot \nabla' \times \nabla' \times \mathbf{G}_1(\mathbf{r}' - \mathbf{r}) \right] dV'$$
$$= \oint_S \left\{ [\hat{\mathbf{n}} \times \mathbf{E}_1(\mathbf{r}')] \cdot [\nabla' \times \mathbf{G}_1(\mathbf{r}' - \mathbf{r})] + \hat{\mathbf{n}} \times \nabla' \times \mathbf{E}_1(\mathbf{r}') \cdot \mathbf{G}_1(\mathbf{r}' - \mathbf{r}) \right\} dS'. \quad (4.15)$$

Noting that (see Equation [4.12])

$$\left[\nabla' \times \mathbf{G}_1(\mathbf{r}' - \mathbf{r}) \right]^t = -\nabla' \times \mathbf{G}_1(\mathbf{r} - \mathbf{r}') = \nabla \times \mathbf{G}_1(\mathbf{r} - \mathbf{r}'), \quad (4.16)$$

Equation (4.13) transforms as follows:

$$\mathbf{E}_1(\mathbf{r}) = \mathbf{E}^{(i)}(\mathbf{r}) + \oint_S \left\{ i\omega\mu_1 \mathbf{G}_1(\mathbf{r} - \mathbf{r}') \cdot [\hat{\mathbf{n}} \times \mathbf{H}_1(\mathbf{r}')] \right.$$
$$\left. - [\nabla \times \mathbf{G}_1(\mathbf{r} - \mathbf{r}')] \cdot [\hat{\mathbf{n}} \times \mathbf{E}_1(\mathbf{r})] \right\} dS', \quad (4.17)$$

where the first of Equations (4.4) have been accounted for, and the integral over S_∞ has been canceled in view of the radiation conditions at infinity. Change of the direction of the unit normal $\hat{\mathbf{n}}$ (see Figure 4.1) leads to the first relevant equation:

$$\mathbf{E}_1(\mathbf{r}) = \mathbf{E}^{(i)}(\mathbf{r}) + \int_S \left\{ -i\omega\mu_1 \mathbf{G}_1(\mathbf{r} - \mathbf{r}') \cdot [\hat{\mathbf{n}}(\mathbf{r}') \times \mathbf{H}_1(\mathbf{r}')] \right.$$
$$\left. + [\nabla \times \mathbf{G}_1(\mathbf{r} - \mathbf{r}')] \cdot [\hat{\mathbf{n}}(\mathbf{r}') \times \mathbf{E}_1(\mathbf{r}')] \right\} dS', \forall \mathbf{r} \in V_1. \quad (4.18)$$

In conclusion, the field in the volume V_1 is represented as superposition of that radiated by real and equivalent sources in an unbounded medium of parameters ε_1, μ_1 coincident with those of the volume V_1. The equivalent sources are electric, $\hat{\mathbf{n}} \times \mathbf{H}$, and magnetic, $-\hat{\mathbf{n}} \times \mathbf{E}$, surface currents over S, whose associate fields are computed via the appropriate Green's functions $-i\omega\mu_1 \mathbf{G}_1(\mathbf{r} - \mathbf{r}')$ and $-\nabla \times \mathbf{G}_1(\mathbf{r} - \mathbf{r}')$, respectively.

The second relevant equation that holds $\forall \mathbf{r} \in V_2$ is similar to Equation (4.18), the only differences being that integration is extended to the

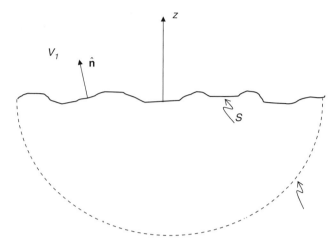

FIGURE 4.3 Relevant to the application of the dyadic divergence theorem to the volume V_2. The region V_2 is closed by means of the surfaces S and S_∞. The usual normal is depicted at a generic point of the surface S.

volume V_2 (see Figure 4.3) and $\mathbf{E}_1 = 0$ because $\mathbf{r}$ is outside V_1:

$$0 = \mathbf{E}^{(i)}(\mathbf{r}) + \int_S \{-i\omega\mu_1 \mathbf{G}_1(\mathbf{r} - \mathbf{r}') \cdot [\hat{\mathbf{n}}(\mathbf{r}') \times \mathbf{H}_1(\mathbf{r}')]$$
$$+ [\nabla \times \mathbf{G}_1(\mathbf{r} - \mathbf{r}')] \cdot [\hat{\mathbf{n}}(\mathbf{r}') \times \mathbf{E}_1(\mathbf{r}')]\} \, dS', \quad \forall \mathbf{r} \in V_2, \quad (4.19)$$

where the volume V_2 is filled with the same material as that of the volume V_1.

Similar derivations can be developed starting from Equations (4.7) and (4.9) instead of Equations (4.6) and (4.8). The final result provides the third and fourth relevant equations:

$$0 = \int_S \{i\omega\mu_2 \mathbf{G}_2(\mathbf{r} - \mathbf{r}') \cdot [\hat{\mathbf{n}}(\mathbf{r}') \times \mathbf{H}_2(\mathbf{r}')]$$
$$- [\nabla \times \mathbf{G}_2(\mathbf{r} - \mathbf{r}')] \cdot [\hat{\mathbf{n}}(\mathbf{r}') \times \mathbf{E}_2(\mathbf{r}')]\} \, dS', \quad \forall \mathbf{r} \in V_1, \quad (4.20)$$

$$\mathbf{E}_2(\mathbf{r}) = \int_S \{i\omega\mu_2 \mathbf{G}_2(\mathbf{r} - \mathbf{r}') \cdot [\hat{\mathbf{n}}(\mathbf{r}') \times \mathbf{H}_2(\mathbf{r}')]$$
$$- [\nabla \times \mathbf{G}_2(\mathbf{r} - \mathbf{r}')] \cdot [\hat{\mathbf{n}}(\mathbf{r}') \times \mathbf{E}_2(\mathbf{r}')]\} \, dS', \quad \forall \mathbf{r} \in V_2, \quad (4.21)$$

where now both volumes V_1 and V_2 are filled with the material of the volume V_2.

4.3. The Integral-Equation Method

Physical implications of Equations (4.18) through (4.21) are now in order.

Equation (4.18) provides the field $\mathbf{E}_1(\mathbf{r})$, solution of Maxwell equations in the volume V_1, in the presence of the source field $\mathbf{E}^{(i)}(\mathbf{r})$, verifying radiation condition at infinity and subject to prescribed continuity conditions of the tangential components $\hat{\mathbf{n}} \times \mathbf{E}_1(\mathbf{r}')$, $\hat{\mathbf{n}} \times \mathbf{H}_1(\mathbf{r}')$ of the field on the boundary surface S. Similarly, Equation (4.21) provides the field, solution of Maxwell equations in the volume V_2, verifying radiation condition at infinity and subject to prescribed continuity conditions of the tangential components $\hat{\mathbf{n}} \times \mathbf{E}_2(\mathbf{r}')$, $\hat{\mathbf{n}} \times \mathbf{H}_2(\mathbf{r}')$ of the field on the boundary surface S.

When the continuity conditions

$$\hat{\mathbf{n}} \times \mathbf{E}_1(\mathbf{r}') = \hat{\mathbf{n}} \times \mathbf{E}_2(\mathbf{r}') = \hat{\mathbf{n}} \times \mathbf{E}(\mathbf{r}'),$$
$$\hat{\mathbf{n}} \times \mathbf{H}_1(\mathbf{r}') = \hat{\mathbf{n}} \times \mathbf{H}_2(\mathbf{r}') = \hat{\mathbf{n}} \times \mathbf{H}(\mathbf{r}'),$$
(4.22)

on S are enforced, then the uniqueness theorem assures that the field $(\mathbf{E}_1, \mathbf{H}_1)$, $(\mathbf{E}_2, \mathbf{H}_2)$ in the volumes V_1, V_2, respectively, are solutions to the problem at hand, verifying Maxwell equations, radiation condition at infinity, and continuity of the tangential components of the fields at the boundary surface S. Their computation is provided by Equations (4.18) and (4.21), once the surface fields on S are known. As already stated, these surface fields are not independent: they must verify Maxwell equations—that is, they are specified, in principle, by the integral Equations (4.19) and (4.20). These are usually referred to as the *electric-field integral equations*; by duality, the corresponding *magnetic-field integral equations* could be obtained.

In conclusion, Equations (4.19) and (4.20) are Fredholm integral equations that define and allow, together with Equations (4.22), computation of the surface tangential fields; these surface fields are used in Equations (4.18) and (4.21), which allow evaluation of the scattered

$$\mathbf{E}^{(s)}(\mathbf{r}) = \mathbf{E}_1(\mathbf{r}) - \mathbf{E}^{(i)}(\mathbf{r}), \forall \mathbf{r} \in V_1, \qquad (4.23)$$

and the transmitted

$$\mathbf{E}^{(t)}(\mathbf{r}) = \mathbf{E}_2(\mathbf{r}), \forall \mathbf{r} \in V_2, \qquad (4.24)$$

fields, respectively.

Equations (4.18) through (4.21) have been mathematically derived starting from Maxwell equations and using some theorems of dyadic calculus. But they can also be interpreted in terms of the equivalence theorem.

First, let all the space be filled with electromagnetic material of parameters ε_1, μ_1. A (electric and magnetic) surface current distribution over S is imposed, such that the field is equal to zero in the lower medium and coincides with the actual field in the upper one. Application of the equivalence theorem leads to Equations (4.18) and (4.19), where $\mathbf{E}^{(i)}(\mathbf{r})$ is the *incident field* (i.e., by definition, the *unperturbed field*): the scattering surface must be removed, and the lower medium must be made coincident with the upper one. If all the space is now filled with a material of electromagnetic parameters coincident with those of the lower medium, application of the equivalence theorem leads to Equations (4.20) and (4.21).

Solution of the system formed by Equations (4,19), (4.20), and (4.22) encounters several difficulties: Fredholm integral equations of the first kind are in order because the unknowns appear only inside the integrals. Closed-form solutions for the tangential fields over the surface are known only in some very limited special cases, the most popular one being the planar scattering surface. Approximate solutions are then required and are introduced in this chapter. Once the electromagnetic field $\mathbf{E}, \mathbf{H}$ at each point $\mathbf{r}'$ of the surface is either known, postulated, measured, or somehow evaluated, the associate scattered field $\mathbf{E}^{(s)}$ at any observation point $\mathbf{r} \in V_1$ is given by

$$\mathbf{E}^{(s)}(\mathbf{r}) = \oint_S \{-i\omega\mu_1 \mathbf{G}_1(\mathbf{r}-\mathbf{r}') \cdot [\hat{\mathbf{n}} \times \mathbf{H}(\mathbf{r}')]$$
$$+ [\nabla \times \mathbf{G}_1(\mathbf{r}-\mathbf{r}')] \cdot [\hat{\mathbf{n}} \times \mathbf{E}_1(\mathbf{r})]\} dS', \quad (4.25)$$

and similarly for the volume V_2.

If the observation point $\mathbf{r}$ is in the far-field region, then the dyadic Green function simplifies as

$$\mathbf{G}_1(\mathbf{r}, \mathbf{r}') = \left[\mathbf{I} - \hat{\mathbf{k}}_s \hat{\mathbf{k}}_s\right] \frac{\exp(-ikr)}{4\pi r} \exp(i\mathbf{k}_s \cdot \mathbf{r}'), \quad (4.26)$$

wherein the unit vector $\hat{\mathbf{k}}_s$ defines the observation direction (see the next section). Hence, the scattered far field along the observation direction $\hat{\mathbf{k}}_s$

can be obtained by substituting Equation (4.26) in Equation (4.25):

$$\mathbf{E}^{(s)}(\mathbf{r}) = -\frac{ik\exp(-ikr)}{4\pi r}(\mathbf{I}-\hat{\mathbf{k}}_s\hat{\mathbf{k}}_s)$$

$$\cdot \int_S \left\{ \hat{\mathbf{k}}_s \times [\hat{\mathbf{n}} \times \mathbf{E}(\mathbf{r}')] + \zeta_1 [\hat{\mathbf{n}} \times \mathbf{H}(\mathbf{r}')] \right\} \exp(i\mathbf{k}_s \cdot \mathbf{r}')dS', \quad (4.27)$$

where $\zeta_1 = \sqrt{\mu_1/\varepsilon_1}$ is the intrinsic impedance of the upper medium.

If expressions obtained in this section are to be applied to a finite part of the surface, as is the case in remote-sensing application, their validity must be judged with respect to the transmitting-antenna and receiving-antenna patterns. As a matter of fact, some considerations could lead to safely applying previous results in case of a surface of finite extent.

4.4. Incident and Scattered-Field Coordinate-Reference Systems

The geometry of the problem is shown in Figure 4.4, wherein a Cartesian (O, x, y, z) and a polar $(O, r, \vartheta, \varphi)$ reference system are introduced; without loss of generality, the $z = 0$ plane is chosen coincident with the natural-surface mean-plane.

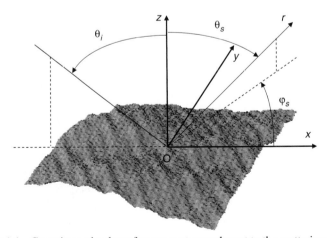

FIGURE 4.4 Cartesian and polar reference systems relevant to the scattering surface.

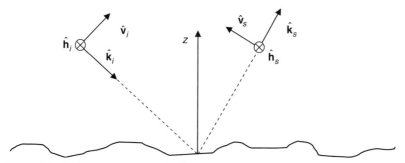

FIGURE 4.5 Incident and scattered reference frames. Incidence and scattering angles are referred to the z-axis, which is orthogonal to the surface mean plane.

Plane-wave superposition can be used to express any incident field: hence, without loss of generality, the incident field is taken to be a single plane wave propagating along the direction $\hat{\mathbf{k}}_i$, identified by the incidence angle ϑ_i over the mean plane and lying in the half plane ($x \leq 0$, $y = 0$) (see Figure 4.5):

$$\begin{cases} \mathbf{E}^{(i)} = \hat{\mathbf{p}} E_p \exp(-i\mathbf{k}_i \cdot \mathbf{r}) \\ \mathbf{H}^{(i)} = \hat{\mathbf{k}}_i \times \mathbf{E}^{(i)}/\zeta_1 \end{cases}$$

$$\begin{cases} \mathbf{k}_i = k\hat{\mathbf{k}}_i = \hat{\mathbf{x}} k \sin \vartheta_i - \hat{\mathbf{z}} k \cos \vartheta_i = \omega\sqrt{\varepsilon_1 \mu_1}\hat{\mathbf{k}}_i \\ \zeta_1 = \sqrt{\mu_1/\varepsilon_1} \end{cases} \quad (4.28)$$

where the unit vector $\hat{\mathbf{p}}$ describes the polarization of the incident field.

It is convenient to introduce the incident-field polarization reference frame $(\hat{\mathbf{h}}_i, \hat{\mathbf{v}}_i, \hat{\mathbf{k}}_i)$:

$$\begin{cases} \hat{\mathbf{h}}_i = \dfrac{\hat{\mathbf{z}} \times \hat{\mathbf{k}}_i}{|\hat{\mathbf{z}} \times \hat{\mathbf{k}}_i|} = \hat{\mathbf{y}} \\ \hat{\mathbf{v}}_i = \hat{\mathbf{k}}_i \times \hat{\mathbf{h}}_i = \hat{\mathbf{x}} \cos \vartheta_i + \hat{\mathbf{z}} \sin \vartheta_i \end{cases} \quad (4.29)$$

where $\hat{\mathbf{z}}$ and $\hat{\mathbf{k}}_i$ define the *mean plane of incidence*, $x = 0$, with respect to the surface mean plane $z = 0$ (see Figure 4.5, specified to the case $\varphi_s = 0$). Then $\hat{\mathbf{p}} = \hat{\mathbf{h}}_i$ orthogonal to the mean plane of incidence calls for horizontal (H, TE, or perpendicular) polarization, whereas the polarization is vertical (V, TM, or parallel) if $\hat{\mathbf{p}} = \hat{\mathbf{v}}_i$.

4.4. Incident and Scattered-Field Coordinate-Reference Systems

Now consider the field scattered in the far-field region along the generic scattering direction $\hat{\mathbf{k}}_s$ identified by the scattering angles ϑ_s and φ_s:

$$\begin{cases} \mathbf{E}^{(s)} = \hat{\mathbf{q}} E_q \exp(-i\mathbf{k}_s \cdot \mathbf{r}) \\ \mathbf{H}^{(s)} = \hat{\mathbf{k}}_s \times \mathbf{E}^{(s)}/\zeta_1 \\ \mathbf{k}_s = k\hat{\mathbf{k}}_s = \hat{\mathbf{x}} k \sin\vartheta_s \cos\varphi_s + \hat{\mathbf{y}} k \sin\vartheta_s \sin\varphi_s + \hat{\mathbf{z}} k \cos\vartheta_s \end{cases}, \quad (4.30)$$

where the unit vector $\hat{\mathbf{q}}$ describes the polarization of the scattered field. As for the incident field, it is convenient to introduce the scattered-field polarization frame $(\hat{\mathbf{h}}_s, \hat{\mathbf{v}}_s, \hat{\mathbf{k}}_s)$:

$$\begin{cases} \hat{\mathbf{h}}_s = \dfrac{\hat{\mathbf{z}} \times \hat{\mathbf{k}}_s}{|\hat{\mathbf{z}} \times \hat{\mathbf{k}}_s|} = -\hat{\mathbf{x}} \sin\varphi_s + \hat{\mathbf{y}} \cos\varphi_s \\ \hat{\mathbf{v}}_s = \hat{\mathbf{k}}_s \times \hat{\mathbf{h}}_s = -\hat{\mathbf{x}} \cos\vartheta_s \cos\varphi_s - \hat{\mathbf{y}} \cos\vartheta_s \sin\varphi_s + \hat{\mathbf{z}} \sin\vartheta_s \end{cases}, \quad (4.31)$$

so that parallel and perpendicular components of the scattered field can be similarly defined.

Incident field and scattered fields can then be expressed in their polarization frames:

$$\begin{cases} \mathbf{E}^{(i)} = \left[(\hat{\mathbf{p}} \cdot \hat{\mathbf{v}}_i)\hat{\mathbf{v}}_i + (\hat{\mathbf{p}} \cdot \hat{\mathbf{h}}_i)\hat{\mathbf{h}}_i\right] E_p \exp(-i\mathbf{k}_i \cdot \mathbf{r}') = E_v^{(i)} \hat{\mathbf{v}}_i + E_h^{(i)} \hat{\mathbf{h}}_i \\ \mathbf{H}^{(i)} = \left[-(\hat{\mathbf{p}} \cdot \hat{\mathbf{v}}_i)\hat{\mathbf{h}}_i + (\hat{\mathbf{p}} \cdot \hat{\mathbf{h}}_i)\hat{\mathbf{v}}_i\right] \dfrac{E_p}{\zeta_1} \exp(-i\mathbf{k}_i \cdot \mathbf{r}') = \dfrac{1}{\zeta_1} \hat{\mathbf{k}}_i \times \mathbf{E}, \end{cases} \quad (4.32)$$

$$\begin{cases} \mathbf{E}^{(s)} = \left[(\hat{\mathbf{p}} \cdot \hat{\mathbf{v}}_s)\hat{\mathbf{v}}_s + (\hat{\mathbf{p}} \cdot \hat{\mathbf{h}}_s)\hat{\mathbf{h}}_s\right] E_q \exp(-i\mathbf{k}_s \cdot \mathbf{r}) = E_v^{(s)} \hat{\mathbf{v}}_s + E_h^{(s)} \hat{\mathbf{h}}_s \\ \mathbf{H}^{(s)} = \left[-(\hat{\mathbf{p}} \cdot \hat{\mathbf{v}}_s)\hat{\mathbf{h}}_s + (\hat{\mathbf{p}} \cdot \hat{\mathbf{h}}_s)\hat{\mathbf{v}}_s\right] \dfrac{E_q}{\zeta_1} \exp(-i\mathbf{k}_s \cdot \mathbf{r}) = \dfrac{1}{\zeta_1} \hat{\mathbf{k}}_s \times \mathbf{E}^{(s)}. \end{cases} \quad (4.33)$$

By definition, the scattering problem consists of computing the scattered field whenever the incident one and the media have been prescribed. In the far-field region, the solution can be formally organized in terms of the *scattering matrix* **S**, relating parallel and perpendicular components of incident and scattered fields:

$$\mathbf{E}^{(s)} = -\dfrac{ik}{4\pi r} \exp(-ikr) \mathbf{S} \cdot \mathbf{E}^{(i)}, \quad (4.34)$$

i.e.,

$$\begin{vmatrix} E_v^{(s)} \\ E_h^{(s)} \end{vmatrix} = -\frac{ik}{4\pi r} \exp(-ikr) \begin{vmatrix} S_{vv} & S_{vh} \\ S_{hv} & S_{hh} \end{vmatrix} \cdot \begin{vmatrix} E_v^{(i)} \\ E_h^{(i)} \end{vmatrix}, \qquad (4.35)$$

where the scalar coefficient before the scattering matrix is a convenient normalization factor.

In this chapter, preliminary steps to get the scattering matrix are referred to.

4.5. The Kirchhoff Approximation

The Kirchhoff approximation provides an estimate of the scattered field tangent to the surface in terms of the incident one: the scattered tangential field at each point of the surface is evaluated by locally approximating the surface with its tangent plane. For each point, the *local incidence angle* is evaluated with reference to the local tangential plane; hence, for rough surfaces, the tangential plane changes over the surface according to the local normal $\hat{\mathbf{n}}$. Thus, KA provides a dramatic simplification in the integral-equation approach: the surface integral equations are no longer necessary, because, in this framework, the surface fields in Equations (4.18) and (4.21) are evaluated with reference to the locally tangent planes: whenever a plane wave is incident over a plane discontinuity, the Fresnel coefficients directly relate, over the discontinuity, scattered and incident tangential fields.

Evaluation of the tangential field over a rough surface under KA is now in order.

It is convenient to introduce for each point of the scattering surface the *local* incident-field-polarization reference frame $(\hat{\mathbf{h}}_l, \hat{\mathbf{v}}_l, \hat{\mathbf{k}}_i)$ (see Figure 4.6):

$$\begin{cases} \hat{\mathbf{h}}_l = \dfrac{\hat{\mathbf{n}} \times \hat{\mathbf{k}}_i}{\left|\hat{\mathbf{n}} \times \hat{\mathbf{k}}_i\right|}, \\ \hat{\mathbf{v}}_l = \hat{\mathbf{k}}_i \times \hat{\mathbf{h}}_l \end{cases} \qquad (4.36)$$

where $\hat{\mathbf{n}}$ and $\hat{\mathbf{k}}_i$ define the local plane of incidence. These unit vectors are defined similarly to those in Section 4.4, Equations (4.29), but they coincide with them only if $\hat{\mathbf{n}} \equiv \hat{\mathbf{z}}$. It is noted that $(\hat{\mathbf{h}}_i, \hat{\mathbf{v}}_i)$ are constant

4.5. The Kirchhoff Approximation

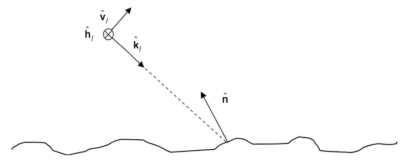

FIGURE 4.6 Local reference frame. Incidence and scattering angles are referred to the normal to the surface, which is orthogonal to the local surface tangent plane.

unit vectors, at variance of $(\hat{\mathbf{h}}_l, \hat{\mathbf{v}}_l)$ which are space dependent over the surface.

The electromagnetic field incident over the surface can now be decomposed onto its local perpendicular and parallel components:

$$\mathbf{E}^{(i)} = \left[(\hat{\mathbf{p}} \cdot \hat{\mathbf{v}}_l)\hat{\mathbf{v}}_l + (\hat{\mathbf{p}} \cdot \hat{\mathbf{h}}_l)\hat{\mathbf{h}}_l\right] E_p \exp(-i\mathbf{k}_i \cdot \mathbf{r}')$$

$$\mathbf{H}^{(i)} = \frac{1}{\zeta_1}\hat{\mathbf{k}}_i \times \mathbf{E}^{(i)} = \left[-(\hat{\mathbf{p}} \cdot \hat{\mathbf{v}}_l)\hat{\mathbf{h}}_l + (\hat{\mathbf{p}} \cdot \hat{\mathbf{h}}_l)\hat{\mathbf{v}}_l\right] \frac{E_p}{\zeta_1} \exp(-i\mathbf{k}_i \cdot \mathbf{r}'),$$
(4.37)

which formally coincide with Equations (4.32), but for the new local reference frame.

To proceed further, the local Fresnel reflection coefficients for perpendicular and parallel polarization are needed.

Two media of (possibly complex) permittivity and permeability ε_1, ε_2, μ_1, μ_2, respectively, separated by the surface S, are considered. The field is incident from the upper medium. The Fresnel reflection coefficients for the two polarizations are

$$R_h = \frac{\mu_2\sqrt{\varepsilon_1\mu_1}\cos\theta_l - \mu_1\sqrt{\varepsilon_2\mu_2 - \varepsilon_1\mu_1\sin^2\theta_l}}{\mu_2\sqrt{\varepsilon_1\mu_1}\cos\theta_l + \mu_1\sqrt{\varepsilon_2\mu_2 - \varepsilon_1\mu_1\sin^2\theta_l}},$$
(4.38)

$$R_v = \frac{\varepsilon_2\sqrt{\varepsilon_1\mu_1}\cos\theta_l - \varepsilon_1\sqrt{\varepsilon_2\mu_2 - \varepsilon_1\mu_1\sin^2\theta_l}}{\varepsilon_2\sqrt{\varepsilon_1\mu_1}\cos\theta_l + \varepsilon_1\sqrt{\varepsilon_2\mu_2 - \varepsilon_1\mu_1\sin^2\theta_l}},$$
(4.39)

where θ_l is the local incidence angle defined by

$$\cos\theta_l = -\hat{\mathbf{n}} \cdot \hat{\mathbf{k}}_i. \tag{4.40}$$

Fresnel coefficients as given by Equations (4.38) and (4.39) refer to the tangential component of the electric field. To get the corresponding coefficient for the magnetic field, only a minus sign must be added.

The more common case pertinent to natural surfaces is considered—that is, propagation in free space, and incidence on a homogeneous medium whose complex relative permittivity is ε_r and whose permeability coincides with that of the free space. Equations (4.38) and (4.39) simplify:

$$R_h = \frac{\cos\vartheta_l - \sqrt{\varepsilon_r - \sin^2\vartheta_l}}{\cos\vartheta_l + \sqrt{\varepsilon_r - \sin^2\vartheta_l}}, \tag{4.41}$$

$$R_v = \frac{\varepsilon_r \cos\vartheta_l - \sqrt{\varepsilon_r - \sin^2\vartheta_l}}{\varepsilon_r \cos\vartheta_l + \sqrt{\varepsilon_r - \sin^2\vartheta_l}}. \tag{4.42}$$

This is the case from now on.

As already anticipated, the scattered field tangent to the surface is taken proportional to the incident one via the Fresnel reflection coefficients:

$$\begin{aligned}\hat{\mathbf{n}} \times (\mathbf{E}^{(s)} \cdot \hat{\mathbf{h}}_l)\hat{\mathbf{h}}_l &= R_h \hat{\mathbf{n}} \times (\mathbf{E}^{(i)} \cdot \hat{\mathbf{h}}_l)\hat{\mathbf{h}}_l \\ \hat{\mathbf{n}} \times (\mathbf{E}^{(s)} \cdot \hat{\mathbf{v}}_l)\hat{\mathbf{v}}_l &= R_v \hat{\mathbf{n}} \times (\mathbf{E}^{(i)} \cdot \hat{\mathbf{v}}_l)\hat{\mathbf{v}}_l\end{aligned}. \tag{4.43}$$

It is convenient to express the normal unity vector $\hat{\mathbf{n}}$ in terms of the surface slopes α and β, which are directly related to the surface profile:

$$\hat{\mathbf{n}}(\mathbf{r}') = \frac{-\alpha\hat{\mathbf{x}} - \beta\hat{\mathbf{y}} + \hat{\mathbf{z}}}{\sqrt{1 + \alpha^2 + \beta^2}}, \tag{4.44}$$

wherein

$$\alpha = \frac{\partial z'(x', y')}{\partial x'}, \quad \beta = \frac{\partial z'(x', y')}{\partial y'}. \tag{4.45}$$

Hence, enforcing KA, the total field tangent to the surface is expressed in terms of the incident one and of the surface geometric and electromagnetic

4.5. The Kirchhoff Approximation

parameters as

$$\hat{n} \times \mathbf{E}(\mathbf{r}') = \hat{n} \times \mathbf{E}^{(i)}(\mathbf{r}') + \hat{n} \times \mathbf{E}^{(s)}(\mathbf{r}')$$

$$= \left[(\hat{\mathbf{p}} \cdot \hat{\mathbf{v}}_l)(\hat{n} \times \hat{\mathbf{v}}_l)(1 + R_v) + (\hat{\mathbf{p}} \cdot \hat{\mathbf{h}}_l)(\hat{n} \times \hat{\mathbf{h}}_l)(1 + R_h)\right]$$

$$\times E_p \exp(-i\mathbf{k}_i \cdot \mathbf{r}')$$

$$\hat{n} \times \mathbf{H}(\mathbf{r}') = \hat{n} \times \mathbf{H}^{(i)}(\mathbf{r}') + \hat{n} \times \mathbf{H}^{(s)}(\mathbf{r}')$$

$$= \left[(\hat{\mathbf{p}} \cdot \hat{\mathbf{h}}_l)(\hat{n} \times \hat{\mathbf{v}}_l)(1 - R_h) - (\hat{\mathbf{p}} \cdot \hat{\mathbf{v}}_l)(\hat{n} \times \hat{\mathbf{h}}_l)(1 - R_v)\right]$$

$$\times \frac{E_p}{\zeta_0} \exp(-i\mathbf{k}_i \cdot \mathbf{r}'), \tag{4.46}$$

where ζ_0 is the free-space intrinsic impedance.

In the far-field region, Equation (4.27) can be used, and the scattered field can be conveniently expressed as

$$\mathbf{E}^{(s)}(\mathbf{r}) = -\frac{ik \exp(-ikr)}{4\pi r} E_p (I - \hat{\mathbf{k}}_s \hat{\mathbf{k}}_s) \cdot \iint_A \mathbf{F}_p(\alpha, \beta) \exp\left[-i(\mathbf{k}_i - \mathbf{k}_s) \cdot \mathbf{r}'\right] dA', \tag{4.47}$$

where $\mathbf{r} = [x, y, z(x, y)]$, $A = XY$ is the illuminated area S projected onto the (x, y) plane, and

$$\mathbf{F}_p(\alpha, \beta) = \left\{(\hat{\mathbf{p}} \cdot \hat{\mathbf{v}}_l)\left[\hat{\mathbf{k}}_s \times (\hat{n} \times \hat{\mathbf{v}}_l)\right](1 + R_v) + (\hat{\mathbf{p}} \cdot \hat{\mathbf{h}}_l)\left[\hat{\mathbf{k}}_s \times (\hat{n} \times \hat{\mathbf{h}}_l)\right]\right.$$

$$(1 + R_h) + \left[(\hat{\mathbf{p}} \cdot \hat{\mathbf{h}}_l)(\hat{n} \times \hat{\mathbf{v}}_l)(1 - R_h)\right.$$

$$\left.\left. - (\hat{\mathbf{p}} \cdot \hat{\mathbf{v}}_l)(\hat{n} \times \hat{\mathbf{h}}_l)(1 - R_v)\right]\right\} \sqrt{1 + \alpha^2 + \beta^2}. \tag{4.48}$$

In Equation (4.48), the square root accounts for the Jacobian of the transformation of the integration variables from S onto A.

In conclusion, KA allows us to express the scattered field in terms of the incident one and of the surface shape: it is not necessary to solve the surface integral equations, and computation of the scattered field in closed form requires only evaluation of the integral in Equation (4.47). However, this evaluation is not straightforward. As a matter of fact, the vector function, $\mathbf{F}_p(\cdot)$ in Equation (4.48) exhibits an involved dependence on the surface profile; evaluation of the integral in Equation (4.47) can be obtained in closed form only for very limited geometric deterministic shapes of scattering bodies. In the case of rough surfaces, appropriate approximations are explored to get a closed-form solution for the scattered field. Rationale for these approximations is presented in the next section.

4.6. Physical-Optics Solution

Within KA, further approximations lead to closed-form analytic solutions for the scattered field (see Equations [4.47] and [4.48])—namely, Geometric Optics (GO) and Physical Optics (PO).

GO makes use of an asymptotic expansion of the integral in Equation (4.47), valid in the so-called high-frequency regime, whose occurrence in conjunction with fractal models for the scattering surface may be questionable. As a matter of fact, fractal surfaces are rough at any scale, whereas GO solution requires the surface to appear smooth as the electromagnetic frequency increases. Hence, GO solution is not explored here, even if in the following chapters some high-frequency-regime solutions are presented.

Alternatively, PO makes use of a series expansion of the function $\mathbf{F}_p(\alpha, \beta)$ in Equation (4.48) in terms of the surface local slopes α and β:

$$\mathbf{F}_p(\alpha, \beta) = \mathbf{F}_p(0,0) + \left.\frac{\partial \mathbf{F}_p(\alpha, \beta)}{\partial \alpha}\right|_{0,0} \alpha + \left.\frac{\partial \mathbf{F}_p(\alpha, \beta)}{\partial \beta}\right|_{0,0} \beta + \cdots \quad (4.49)$$

The *small-slope-regime* solution is obtained in case only the zero-order term of the series is retained; this term is constant with respect to the surface slope. Differently stated, within the small-slope regime, in Equation (4.49), the normal to the surface becomes coincident with the z-axis—namely, $\hat{\mathbf{n}} \equiv \hat{\mathbf{z}}$. It follows that $\mathbf{F}_p(\cdot)$ becomes independent on the surface corrugations, at variance of the exponential term within the integral in Equation (4.47), where this dependence is retained. This approach is also a preliminary step toward higher-order solutions that can be recovered, from the small-slope-regime one, by retaining higher-order terms in Equation (4.49) and proceeding via integration by parts of the integral in Equation (4.47).

It is convenient to introduce the polarization-dependent scalar component $F_{pq}(\cdot)$:

$$F_{pq}(\vartheta_i, \vartheta_s, \varphi_s) = \hat{\mathbf{q}} \cdot \left[(I - \hat{\mathbf{k}}_s \hat{\mathbf{k}}_s) \cdot \mathbf{F}_p(0,0)\right], \quad (4.50)$$

4.6. Physical-Optics Solution

where the functional dependence on the incident and scattering angles over the mean plane has now been highlighted. This dimensionless scalar function, $F_{pq}(\cdot)$, can be evaluated in the small-slope regime by means of Equation (4.49), with $\alpha = \beta = 0$, $\hat{\mathbf{n}} \equiv \hat{\mathbf{z}}$, leading to simple expressions for $F_{pq}(\cdot)$ that depend on the Fresnel reflection coefficients over the mean plane, on the incidence and scattering angles, and on the polarizations of the incident wave (p) and the receiving antenna (q):

$$F_{hh}(\vartheta_i, \vartheta_s, \varphi_s) = [-(1 + R_{h0})\cos\vartheta_s + (1 - R_{h0})\cos\vartheta_i]\cos\varphi_s$$
$$F_{hv}(\vartheta_i, \vartheta_s, \varphi_s) = [(1 + R_{h0}) - (1 - R_{h0})\cos\vartheta_i \cos\vartheta_s]\sin\varphi_s$$
$$F_{vh}(\vartheta_i, \vartheta_s, \varphi_s) = [(1 + R_{v0})\cos\vartheta_i \cos\vartheta_s - (1 - R_{v0})]\sin\varphi_s$$
$$F_{vv}(\vartheta_i, \vartheta_s, \varphi_s) = [-(1 - R_{v0})\cos\vartheta_s + (1 + R_{v0})\cos\vartheta_i]\cos\varphi_s$$

(4.51)

In Equations (4.51), R_{v0} and R_{h0} are the (parallel and perpendicular) Fresnel reflection coefficients evaluated from Equations (4.41) and (4.42) by setting $\vartheta_l = \vartheta_i$,—that is, by making the incidence angle coincide with that over the mean plane.

It is concluded that in the far-field region, PO solution, in the small-slope regime, leads to the following expressions for the entries of the scattering matrix:

$$S_{pq} = F_{pq}(\vartheta_i, \vartheta_s, \varphi_s) \iint_A \exp(-i\boldsymbol{\eta} \cdot \mathbf{r}')dA', \qquad (4.52)$$

where (p, q) may take any value (h, v), and $\boldsymbol{\eta} = \mathbf{k}_i - \mathbf{k}_s$ is the vector whose components in the reference (x, y, z) plane are given by

$$\begin{cases} \eta_x = k(\sin\vartheta_i - \sin\vartheta_s \cos\varphi_s) \\ \eta_y = -k\sin\vartheta_s \sin\varphi_s \\ \eta_z = -k(\cos\vartheta_i + \cos\vartheta_s) \end{cases} \qquad (4.53)$$

In Equation (4.53), all angles are referred to the scattering mean plane as illustrated in Figure 4.4.

In the small-slope expansion, the surface profile is still included in the exponential term $\mathbf{r}'$ inside the diffraction integral. Hence, getting closed-form solutions for the scattered field requires that the surface geometric shape must be considered in the argument of the exponential inside the integral in Equation (4.47). This is done in Chapter 5 for WM, and in Chapter 6 for fBm fractal surfaces, respectively. In the former case, the surface is described by a predictable stochastic process, and the scattered field is evaluated as a predictable stochastic process as well; in the latter case, the surface is described by a regular stochastic process, and the scattered field is evaluated as a regular stochastic process. For classical surfaces, the only viable closed-form solution can be obtained for the regular stochastic-process description of the surface shape; hence, comparison between the scattered field evaluated starting from classical- and fractal-surface models is presented only in Chapter 6, where regular stochastic processes are accounted for.

4.7. Extended-Boundary-Condition Method

A different approach to addressing the surface integral equation is provided by the Extended-Boundary-Condition Method (EBCM).

Equations (4.19) and (4.20) are now examined, where the equivalent sources $\hat{n} \times \mathbf{H}_1 = \hat{n} \times \mathbf{H}_2 = \hat{n} \times \mathbf{H}, \hat{n} \times \mathbf{E}_1 = \hat{n} \times \mathbf{E}_2 = \hat{n} \times \mathbf{E}$, are present, $(\mathbf{E}, \mathbf{H})$ being the (unknown) actual fields over the surface S separating the two media (see Equations [4.22]). It is convenient to introduce the vector unknowns

$$\mathbf{a}(\mathbf{r}'_\perp)d\mathbf{r}'_\perp \stackrel{\Delta}{=} \zeta_1 \hat{n}(\mathbf{r}') \times \mathbf{H}_1(\mathbf{r}')dS' = \zeta_1 \hat{n}(\mathbf{r}') \times \mathbf{H}_2(\mathbf{r}')dS'$$

$$\mathbf{b}(\mathbf{r}'_\perp)d\mathbf{r}'_\perp \stackrel{\Delta}{=} \hat{n}(\mathbf{r}') \times \mathbf{E}_1(\mathbf{r}')dS' = \hat{n}(\mathbf{r}') \times \mathbf{E}_2(\mathbf{r}')dS'$$

(4.54)

where $\mathbf{r}'_\perp = x'\hat{\mathbf{x}} + y'\hat{\mathbf{y}}$, and $d\mathbf{r}'_\perp = dxdy$ is the projection of dS' orthogonal to the z-axis, so that the Jacobian of the transformation from S to the surface mean plane is accounted for. These unknowns are related to the magnetic and electric tangential fields, respectively, are appropriately normalized, and take into account the Jacobian of the transformation from the surface integral in dS' to the double integral in $d\mathbf{r}'_\perp$. It also turns out that $\mathbf{a}$ and $\mathbf{b}$ are both measured in V·m.

Substitution of Equations (4.54) into Equations (4.19) and (4.20) leads to

$$\begin{cases} 0 = \mathbf{E}^{(i)}(\mathbf{r}) + \int_S \{-ik_1 G_1(\mathbf{r}-\mathbf{r}') \cdot \mathbf{a}(\mathbf{r}') \\ \qquad + [\nabla \times G_1(\mathbf{r}-\mathbf{r}')] \cdot \mathbf{b}(\mathbf{r}')\} d\mathbf{r}'_\perp \quad \forall \mathbf{r} \in V_2 \\ 0 = \int_S \{ik_2 G_2(\mathbf{r}-\mathbf{r}') \cdot \mathbf{a}(\mathbf{r}') - [\nabla \times G_2(\mathbf{r}-\mathbf{r}')] \cdot \mathbf{b}(\mathbf{r}')\} d\mathbf{r}'_\perp \quad \forall \mathbf{r} \in V_1 \end{cases}$$

(4.55)

4.7. Extended-Boundary-Condition Method

and substitution into Equations (4.18) and (4.21) leads to

$$\begin{cases} \mathbf{E}_1(\mathbf{r}) = \mathbf{E}^{(i)}(\mathbf{r}) + \int_S \{-ik_1 \mathbf{G}_1(\mathbf{r}-\mathbf{r}') \cdot \mathbf{a}(\mathbf{r}') \\ \qquad + [\nabla \times \mathbf{G}_1(\mathbf{r}-\mathbf{r}')] \cdot \mathbf{b}(\mathbf{r}')\} dr'_\perp \quad \forall \mathbf{r} \in V_1 \\ \mathbf{E}_2(\mathbf{r}) = \int_S \{ik_2 \mathbf{G}_2(\mathbf{r}-\mathbf{r}') \cdot \mathbf{a}(\mathbf{r}') - [\nabla \times \mathbf{G}_2(\mathbf{r}-\mathbf{r}')] \cdot \mathbf{b}(\mathbf{r}')\} dr'_\perp \quad \forall \mathbf{r} \in V_2 \end{cases} \quad (4.56)$$

Equations (4.55) form a system of two integral equations to determine the unknown vectors $\mathbf{a}$, $\mathbf{b}$; then Equations (4.56) are used to compute the scattered and, if required, the transmitted fields. A further manipulation of Equations (4.55) and (4.56) is convenient to render the solution of the system more easy and to facilitate straightforward computation of the scattered field.

Up to this point, two Green's dyadics have been introduced: $\mathbf{G}_1(\cdot)$ and $\mathbf{G}_2(\cdot)$, where the lower index specifies the (unbounded) medium, (ε_1, μ_1) and (ε_2, μ_2), where the dyadics are defined. It is now convenient to represent these dyadics by means of their spectral form. In this case, a single expression for each dyadic is not available, and it is convenient to introduce different Green's dyadics individuated by an additional upper index $\mathbf{G}_1^+(\cdot)$, $\mathbf{G}_1^-(\cdot)$, $\mathbf{G}_2^+(\cdot)$ and $\mathbf{G}_2^-(\cdot)$. At variance of the previous ones, these dyadics hold spectral representations that satisfy radiation condition at $z \to +\infty$, upper index plus, or $z \to -\infty$, upper index minus. The reason for this choice is the convenience of expressing the Green's functions by means of their plane-wave expansion:

$$\mathbf{G}_1^\pm(\mathbf{r}-\mathbf{r}') = -\hat{\mathbf{z}}\hat{\mathbf{z}}\frac{\delta(\mathbf{r}-\mathbf{r}')}{k_1^2} - \frac{i}{8\pi^2}\int_{-\infty}^{+\infty} d\mathbf{k}_\perp \frac{1}{k_{1z}}\left[\hat{\mathbf{v}}_1^\pm \hat{\mathbf{v}}_1^\pm + \hat{\mathbf{h}}_1^\pm \hat{\mathbf{h}}_1^\pm\right]$$
$$\exp\left[-i(\mathbf{k}_\perp \pm k_{1z}\hat{\mathbf{z}}) \cdot (\mathbf{r}-\mathbf{r}')\right], \quad (4.57)$$

$$\mathbf{G}_2^\pm(\mathbf{r}-\mathbf{r}') = -\hat{\mathbf{z}}\hat{\mathbf{z}}\frac{\delta(\mathbf{r}-\mathbf{r}')}{k_2^2} - \frac{i}{8\pi^2}\int_{-\infty}^{+\infty} d\mathbf{k}_\perp \frac{1}{k_{2z}}\left[\hat{\mathbf{v}}_1^\pm \hat{\mathbf{v}}_1^\pm + \hat{\mathbf{h}}_1^\pm \hat{\mathbf{h}}_1^\pm\right]$$
$$\exp\left[-i(\mathbf{k}_\perp \pm k_{2z}\hat{\mathbf{z}}) \cdot (\mathbf{r}-\mathbf{r}')\right], \quad (4.58)$$

where $\mathbf{k}_{1,2} = k_x\hat{\mathbf{x}} + k_y\hat{\mathbf{y}} + k_{(1,2)z}\hat{\mathbf{z}}$, $k_{1,2}^2 = k_x^2 + k_y^2 + k_{(1,2)z}^2$ and $\mathbf{k}_\perp = k_x\hat{\mathbf{x}} + k_y\hat{\mathbf{y}}$. In Equations (4.57) and (4.58), the other unit vectors

$$\begin{cases} \hat{\mathbf{h}}^+ = \dfrac{\hat{\mathbf{z}} \times \hat{\mathbf{k}}}{|\hat{\mathbf{z}} \times \hat{\mathbf{k}}|} \\ \hat{\mathbf{v}}^+ = \hat{\mathbf{k}} \times \hat{\mathbf{h}}^+ \end{cases} \quad (4.59)$$

and

$$\begin{cases} \hat{\mathbf{h}}^- = \hat{\mathbf{h}}^+ \\ \hat{\mathbf{v}}^- = \dfrac{(\mathbf{k}_\perp - k_z\hat{\mathbf{z}})}{k} \times \hat{\mathbf{h}}^- \end{cases} \quad (4.60)$$

are present, and define the orthonormal systems $\left(\hat{\mathbf{h}}^+, \hat{\mathbf{v}}^+, \dfrac{(\mathbf{k}_\perp + k_z\hat{\mathbf{z}})}{k}\right)$ and $\left(\hat{\mathbf{h}}^-, \hat{\mathbf{v}}^-, \dfrac{(\mathbf{k}_\perp - k_z\hat{\mathbf{z}})}{k}\right)$ with $I = \dfrac{(\mathbf{k}_\perp + k_z\hat{\mathbf{z}})}{k}\dfrac{(\mathbf{k}_\perp + k_z\hat{\mathbf{z}})}{k} + \hat{\mathbf{v}}^+\hat{\mathbf{v}}^+ + \hat{\mathbf{h}}^+\hat{\mathbf{h}}^+ = \dfrac{(\mathbf{k}_\perp - k_z\hat{\mathbf{z}})}{k}\dfrac{(\mathbf{k}_\perp - k_z\hat{\mathbf{z}})}{k} + \hat{\mathbf{v}}^-\hat{\mathbf{v}}^- + \hat{\mathbf{h}}^-\hat{\mathbf{h}}^-$. Equations (4.57) and (4.58) provide fields radiated by elementary sources that are located on the scattering surface and radiate in an unbounded homogeneous medium toward $z \to \pm\infty$, depending on the choice of their upper and lower indexes. The total field is expressed in terms of a continuous superposition of horizontally and vertically polarized (with respect to the mean surface plane, $z = 0$) plane waves whose wavevectors are provided by $\mathbf{k}$.

Use of the Green's functions in Equations (4.55) and (4.56) requires specification of their indexes. But a problem must be faced (see Figure 4.7), depending on the position of point $\mathbf{r}$. Let z_M, z_m be the appropriate coordinate along the z-axis such that all the shape of the surface is included within them. Then, specification of Equations (4.55) to $\mathbf{r}(Q_2)$ requires use of only the Green's function $G_1^-(\cdot)$, where reference is made to the first of Equations (4.55), because Q_2 lies in V_2. And similarly for $\mathbf{r}(Q_1)$, where the

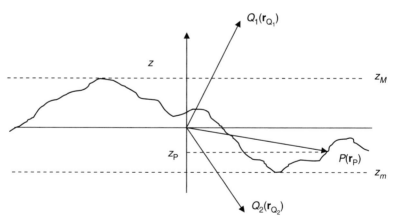

FIGURE 4.7 Relevant to the use of the Green's function in the EBCM. Homogeneous regions are defined by $z < z_m$ and $z > z_M$.

4.7. Extended-Boundary-Condition Method

appropriate function is $G_2^+(\cdot)$. This is at variance of point $\mathbf{r}(P)$, where $G_1^+(\cdot)$ must be used for $z' < z_p$, and $G_1^-(\cdot)$ for $z' > z_p$. This is very inconvenient for subsequent manipulation of Equations (4.55) and (4.56).

Use of the EBCM implies an approximation: to enforce Equations (4.55)—not over the full upper and lower spaces, respectively, but only for $z > z_M$ and $z < z_m$. Accordingly, it is not guaranteed that the equations are verified in the strip between the coordinates (z_M, z_m).

Substitution of Equations (4.57) and (4.58) into Equations (4.55) leads to

$$\mathbf{E}^{(i)}(\mathbf{r}) = \frac{1}{8\pi^2} \int_{-\infty}^{+\infty} d\mathbf{k}_\perp \exp\left[-i(\mathbf{k}_\perp - k_{1z}\hat{\mathbf{z}}) \cdot \mathbf{r}\right]$$

$$\frac{k_1}{k_{1z}} \int_{-\infty}^{+\infty} d\mathbf{r}'_\perp \exp\left[i(\mathbf{k}_\perp - k_{1z}\hat{\mathbf{z}}) \cdot \mathbf{r}'\right]$$

$$\left\{\left[\hat{\mathbf{v}}_1^- \hat{\mathbf{v}}_1^- + \hat{\mathbf{h}}_1^- \hat{\mathbf{h}}_1^-\right] \cdot \mathbf{a}(\mathbf{r}'_\perp) + \left[\hat{\mathbf{h}}_1^- \hat{\mathbf{v}}_1^- - \hat{\mathbf{v}}_1^- \hat{\mathbf{h}}_1^-\right] \cdot \mathbf{b}(\mathbf{r}'_\perp)\right\}, \quad \forall \mathbf{r} \in V_2, \tag{4.61}$$

$$0 = \frac{1}{8\pi^2} \int_{-\infty}^{+\infty} d\mathbf{k}_\perp \exp\left[-i(\mathbf{k}_\perp + k_{2z}\hat{\mathbf{z}}) \cdot \mathbf{r}\right]$$

$$\frac{k_2}{k_{2z}} \int_{-\infty}^{+\infty} d\mathbf{r}'_\perp \exp\left[i(\mathbf{k}_\perp + k_{2z}\hat{\mathbf{z}}) \cdot \mathbf{r}'\right]$$

$$\left\{\frac{k_1}{k_2}\left[\hat{\mathbf{v}}_2^+ \hat{\mathbf{v}}_2^+ + \hat{\mathbf{h}}_2^+ \hat{\mathbf{h}}_2^+\right] \cdot \mathbf{a}(\mathbf{r}'_\perp) + \left[\hat{\mathbf{h}}_2^+ \hat{\mathbf{v}}_2^+ - \hat{\mathbf{v}}_2^+ \hat{\mathbf{h}}_2^+\right] \cdot \mathbf{b}(\mathbf{r}'_\perp)\right\}, \quad \forall \mathbf{r} \in V_1. \tag{4.62}$$

Equations (4.61) and (4.62) must be solved in terms of the surface tangential fields, $\mathbf{a}$ and $\mathbf{b}$. Then it is possible to express the scattered field in the upper medium:

$$\mathbf{E}^{(s)}(\mathbf{r}) = \mathbf{E}(\mathbf{r}) - \mathbf{E}^{(i)}(\mathbf{r})$$

$$= -\frac{1}{8\pi^2} \int_{-\infty}^{+\infty} d^2\mathbf{k}_\perp \exp\left[-i(\mathbf{k}_\perp + k_{1z}\hat{\mathbf{z}}) \cdot \mathbf{r}\right]$$

$$\frac{k_1}{k_{1z}} \int_{-\infty}^{+\infty} d^2\mathbf{r}'_\perp \exp\left[i(\mathbf{k}_\perp + k_{1z}\hat{\mathbf{z}}) \cdot \mathbf{r}'\right]$$

$$\left\{\left[\hat{\mathbf{v}}_1^+ \hat{\mathbf{v}}_1^+ + \hat{\mathbf{h}}_1^+ \hat{\mathbf{h}}_1^+\right] \cdot \mathbf{a}(\mathbf{r}'_\perp) + \left[\hat{\mathbf{h}}_1^- \hat{\mathbf{v}}_1^- - \hat{\mathbf{v}}_1^- \hat{\mathbf{h}}_1^-\right] \cdot \mathbf{b}(\mathbf{r}'_\perp)\right\} \tag{4.63}$$

and the transmitted field in the lower medium:

$$\mathbf{E}^{(t)}(\mathbf{r}) = \frac{1}{8\pi^2} \int_{-\infty}^{+\infty} d^2\mathbf{k}_\perp \exp\left[-i(\mathbf{k}_\perp - k_{2z}\hat{\mathbf{z}}) \cdot \mathbf{r}\right]$$

$$\frac{k_2}{k_{2z}} \int_{-\infty}^{+\infty} d^2\mathbf{r}'_\perp \exp\left[i(\mathbf{k}_\perp - k_{2z}\hat{\mathbf{z}}) \cdot \mathbf{r}'\right]$$

$$\left\{ \frac{k_1}{k_2} \left[\hat{\mathbf{v}}_2^- \hat{\mathbf{v}}_2^- + \hat{\mathbf{h}}_2^- \hat{\mathbf{h}}_2^-\right] \cdot \mathbf{a}(\mathbf{r}'_\perp) + \left[\hat{\mathbf{h}}_2^+ \hat{\mathbf{v}}_2^+ - \hat{\mathbf{v}}_2^+ \hat{\mathbf{h}}_2^+\right] \cdot \mathbf{b}(\mathbf{r}'_\perp) \right\},$$
(4.64)

wherein Equations (4.56) have been used.

A closed-form solution for the scattered field can be obtained by appropriately expanding the incident-, scattered-, and unknown-surface fields whose tangential components are **a** and **b** (see Section 4.8). For periodic surfaces, these expansions allow us to get closed-form solution for the scattered-power density. Conversely, in the case of fractal-surface models, the scattered field can be evaluated in an appropriate closed form that requires only numerically solving an algebraic system of equations: this is done in Chapter 7.

4.8. Small-Perturbation Method

A formal evaluation of the equations derived in the EBCM approach is necessary to address the SPM solution.

It is convenient to separate the z-component of the unknowns, **a** and **b**, from the transverse ones:

$$\begin{aligned}\mathbf{a}(\mathbf{r}'_\perp) &= \mathbf{a}_\perp \hat{\mathbf{r}}'_\perp + a_z \hat{\mathbf{z}} \\ \mathbf{b}(\mathbf{r}'_\perp) &= \mathbf{b}_\perp \hat{\mathbf{r}}'_\perp + b_z \hat{\mathbf{z}}\end{aligned} \quad ; \quad (4.65)$$

Unknowns **a** and **b** are both tangent to the rough surface whose (local) normal is $\hat{\mathbf{n}}(\mathbf{r}'_\perp)$ (see Equations [4.54]); then it follows that

$$\begin{aligned}\hat{\mathbf{n}}(\mathbf{r}'_\perp) \cdot \mathbf{a}(\mathbf{r}'_\perp) &= 0 \\ \hat{\mathbf{n}}(\mathbf{r}'_\perp) \cdot \mathbf{b}(\mathbf{r}'_\perp) &= 0\end{aligned} \quad . \quad (4.66)$$

4.8. Small-Perturbation Method

Equations (4.66) state that the unknown z-components can be expressed by means of the transverse ones:

$$a_z(\mathbf{r}'_\perp) = \left(\hat{\mathbf{x}}\frac{\partial z'(\mathbf{r}'_\perp)}{\partial x'} + \hat{\mathbf{y}}\frac{\partial z'(\mathbf{r}'_\perp)}{\partial y'}\right) \cdot \mathbf{a}_\perp(\mathbf{r}'_\perp) \qquad (4.67)$$

$$b_z(\mathbf{r}'_\perp) = \left(\hat{\mathbf{x}}\frac{\partial z'(\mathbf{r}'_\perp)}{\partial x'} + \hat{\mathbf{y}}\frac{\partial z'(\mathbf{r}'_\perp)}{\partial y'}\right) \cdot \mathbf{b}_\perp(\mathbf{r}'_\perp) \qquad (4.68)$$

It is concluded that four scalar unknowns must be evaluated, while the six scalar (integral) Equations (4.61) and (4.62) should be enforced. This apparent paradox is resolved by noting that in each one of the two homogeneous media, $z > z_M$ and $z < z_m$, the z-component of the electric field may be expressed as a function of the transverse one. Accordingly, two independent (vector) equations are obtained by projection of Equations (4.61) and (4.62) onto the transverse plane.

Different techniques to solve Equations (4.54), (4.61), and (4.62) in terms of $\mathbf{a}$ and $\mathbf{b}$ can be employed. Among these techniques, a popular one is the Small-Perturbation Method (SPM), whose rationale is described in this section.

The SPM requires that the exponential terms in the integrals in Equations (4.61) and (4.62) are expanded in power series for small values of $k_z z'$:

$$\exp\left[\pm i k_z z'(\mathbf{r}'_\perp)\right] = \sum_{m=0}^{\infty} \frac{\left[\pm i k_z z'(\mathbf{r}'_\perp)\right]^m}{m!}, \qquad (4.69)$$

$$\exp\left[\pm i k_{1z} z'(\mathbf{r}'_\perp)\right] = \sum_{m=0}^{\infty} \frac{\left[\pm i k_{1z} z'(\mathbf{r}'_\perp)\right]^m}{m!}. \qquad (4.70)$$

Each term of Series (4.69) and (4.70) leads to a corresponding-order solution for the SPM. More explicitly, the m-th term in Equations (4.69) and (4.70) leads to the m-th-order solution. Then the complete solution for the SPM is obtained via an appropriate series expansion of the unknowns $\mathbf{a}$ and $\mathbf{b}$:

$$\mathbf{a}(\mathbf{r}'_\perp) = \sum_{m=0}^{\infty} \mathbf{a}^{(m)}(\mathbf{r}'_\perp), \qquad (4.71)$$

$$\mathbf{b}(\mathbf{r}'_\perp) = \sum_{m=0}^{\infty} \mathbf{b}^{(m)}(\mathbf{r}'_\perp), \qquad (4.72)$$

where $\mathbf{a}^{(m)}$ and $\mathbf{b}^{(m)}$ are the m-th order solution for $\mathbf{a}$ and $\mathbf{b}$, respectively. A discussion about the correct order implementation is in order.

Examination of Equations (4.67) and (4.68) shows that the z-components of $\mathbf{a}$ and $\mathbf{b}$ are of higher order compared to the transverse ones, if the small-slope assumption for the scattering surface is enforced. Accordingly, it is possible to assume that

$$a_z^{(0)}(\mathbf{r}'_\perp) = 0, \quad b_z^{(0)}(\mathbf{r}'_\perp) = 0, \tag{4.73}$$

at the order $m = 0$. By implementing this recursive approach at any order $m \geq 1$, it turns out that

$$a_z^{(m)}(\mathbf{r}'_\perp) = \left(\hat{\mathbf{x}}\frac{\partial z'(\mathbf{r}'_\perp)}{\partial x'} + \hat{\mathbf{y}}\frac{\partial z'(\mathbf{r}'_\perp)}{\partial y'}\right) \cdot \mathbf{a}_\perp^{(m-1)}(\mathbf{r}'_\perp), \tag{4.74}$$

$$b_z^{(m)}(\mathbf{r}'_\perp) = \left(\hat{\mathbf{x}}\frac{\partial z'(\mathbf{r}'_\perp)}{\partial x'} + \hat{\mathbf{y}}\frac{\partial z'(\mathbf{r}'_\perp)}{\partial y'}\right) \cdot \mathbf{b}_\perp^{(m-1)}(\mathbf{r}'_\perp). \tag{4.75}$$

It is important to remark that in the small-slope expansion, the surface profile is still included in the exponential term $\mathbf{r}'$ under the diffraction integral. Hence, to get closed-form solution for the scattered field, the surface geometric shape has to be introduced in Equation (4.63). This is done in Chapter 7 for WM fractal surfaces, and in Chapter 8 for fBm surfaces. In the first case, being the surface described by a regular stochastic process, the scattered field is evaluated as a regular stochastic process; in the second one, being the surface described by a predictable stochastic process, the scattered field is evaluated as a predictable stochastic process. For classical surfaces, the only viable closed-form solution can be obtained for the regular stochastic-process description of the surface shape; hence, comparison with scattered fields evaluated starting from surface fractal models can only be made with the material appearing in Chapter 8.

4.9. References and Further Readings

Scattering from random rough surfaces is the main subject of many books. Fundamentals of the scattering problem are reported in any book quoted in this section. Some relevant peculiarities are underlined here. Fundamentals on scattering theory are pioneered in the work of Beckmann and Spizzichino (1987). A study by Ulaby, Moore, and Fung (1982) presents the scattering theory underlying KA and SPM, along with fundamentals on electromagnetic theory and microwave remote-sensing sensors and applications. Chew (1995) derives the integral equation for the scalar as well as the vectorial case of interest in electromagnetics. The work of Tsang, Kong, and Shin (1985) reports EBCM along with the Rayleigh interpretation. The integral-equation method is illustrated by Fung (1994). In Chew's book, as well as in a study by Ishimaru (1993), the theory for propagation in random media is the main topic.

CHAPTER 5

Scattering from Weierstrass-Mandelbrot Surfaces: Physical-Optics Solution

5.1. Introduction and Chapter Outline

In this chapter, a closed-form PO solution in the small-slope regime under KA is obtained for the electromagnetic field scattered by natural surfaces modeled by means of the WM function.

First, the analytic derivation of the scattered field in closed form is systematically presented in Section 5.2: the main result is that the diffraction integral defined in Chapter 4 can be evaluated in closed form whenever the WM function is used to model the geometric features of the scattering surface. The final formula for the scattered field deserves full attention for understanding the physical meaning of its terms: this is done in Section 5.3, showing that the obtained scattered-field structure can be linked to the outcomes of the Floquet theory on scattering from periodic or almost-periodic surfaces. Limits of validity of this scattered-field solution are presented in Section 5.4. Dependence of the scattered field on the WM model and electromagnetic parameters is illustrated in Section 5.5. A comment on the statistics of the scattered field is included in Section 5.6. Key references and suggestions for further readings are reported in Section 5.7.

5.2. Analytic Derivation of the Scattered Field

Analytic evaluation of the scattered field (see Equation [4.47]) in a closed form may be obtained in some cases if an analytic expression for the scattering surface is provided. This is the case for planar or sinusoidal surfaces and for a few other canonical surfaces; no closed-form expression is available for the field scattered by a natural surface modeled by means of classical geometry, even if an analytic expression for the surface should be somehow provided. In this chapter, natural surfaces are modeled by means of the WM function, as shown in Chapter 3; if the natural surface is random, then random coefficients for the WM function should be used.

As already presented in Chapter 3, let us represent the surface $z(x, y)$ geometric properties via a WM function:

$$z(x,y) = B \sum_{p=0}^{P-1} C_p v^{-Hp} \sin\left[\kappa_0 v^p (x \cos \Psi_p + y \sin \Psi_p) + \Phi_p\right], \quad (5.1)$$

wherein B [m] is the overall amplitude-scaling factor; P is the number of tones; κ_0 [m^{-1}] is the wave number of the surface fundamental tone (corresponding to $p = 0$); v (greater than 1) is the seed of the geometric progression that accounts for spectral separation of consecutive tones of the surface; $0 < H < 1$ is the Hurst exponent; and C_p, Ψ_p, Φ_p are deterministic or random coefficients that account, respectively, for amplitude, direction, and phase of each tone.

As shown in Chapter 4, the PO solution to the KA provides the scattered far field as

$$E_{pq}^{(s)}(\mathbf{r}) = -\frac{ik \exp(-ikr)}{4\pi r} E_p F_{pq} \iint_A \exp\left[-i(\mathbf{k}_i - \mathbf{k}_s) \cdot \mathbf{r}'\right] dA, \quad (5.2)$$

where p and q stand for transmitting and receiving polarizations of the electric field, usually taken as horizontal and vertical. The reference system is depicted in Figure 5.1. From now on, the prime that should appear inside the integral of Equation (5.2) is omitted for the sake of simplicity. Substitution of Equation (5.1) in Equation (5.2) leads to

$$E_{pq}^{(s)}(\mathbf{r}) = -\frac{ik \exp(-ikr)}{4\pi r} E_p F_{pq}$$

$$\int_A \exp\left\{i \sum_{p=0}^{P-1} u_p \sin\left[\Omega_p(x,y) + \Phi_p\right]\right\} \exp\left\{-i[\eta_x x + \eta_y y]\right\} dA, \quad (5.3)$$

5.2. Analytic Derivation of the Scattered Field

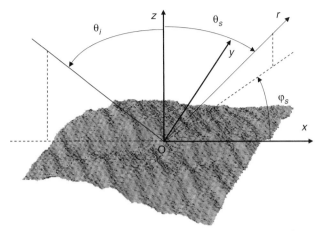

FIGURE 5.1 Geometry of the scattering problem. Cartesian and polar reference systems relevant to the scattering surface are depicted.

wherein $\boldsymbol{\eta} = \mathbf{k}_i - \mathbf{k}_s$ can be expressed in the Cartesian coordinate system as

$$\boldsymbol{\eta} = \mathbf{k}_i - \mathbf{k}_s = \hat{\mathbf{x}}k(\sin\vartheta_i - \sin\vartheta_s \cos\varphi_s) - \hat{\mathbf{y}}k \sin\vartheta_s \sin\varphi_s$$
$$+ \hat{\mathbf{z}}k(\cos\vartheta_i - \cos\vartheta_s) \quad (5.4)$$

(see Chapter 4), and

$$u_p = -\eta_z BC_p v^{-Hp}$$
$$\Omega_p(x, y) = \kappa_0 v^p(x \cos\Psi_p + y \sin\Psi_p). \quad (5.5)$$

Equation (5.3) can be rewritten as

$$E_{pq}^{(s)}(\mathbf{r}) = -\frac{ik \exp(-ikr)}{4\pi r} E_p F_{pq}(\vartheta_i, \vartheta_s, \varphi_s)$$
$$\int_A \exp\{-i[\eta_x x + \eta_y y]\} \prod_{p=0}^{P-1} \exp\{iu_p \sin[\Omega_p(x,y) + \Phi_p]\} dA, \quad (5.6)$$

where the dependence on the integration variables is too involved to allow a closed-form solution for the scattered field. For the exponential terms, use of the Jacobi-Anger expansion, Equation (A.5.1) with $a = u_p$ and $\xi = \Omega_p + \Phi_p$, leads to a more manageable expression of the integrand, because the integration variables in the exponential function appear in linear instead of sinusoidal form:

$$E_{pq}^{(s)}(\mathbf{r}) = -\frac{ik\exp(-ikr)}{4\pi r}E_pF_{pq}(\vartheta_i,\vartheta_s,\varphi_s)\int_A \exp\{-i[\eta_x x + \eta_y y]\}$$
$$\prod_{p=0}^{P-1}\sum_{m=-\infty}^{+\infty} J_m(u_p)\exp\{im[\Omega_p(x,y)+\Phi_p]\}\,dA. \tag{5.7}$$

In Equation (5.7), the integral and the product cannot be exchanged because Ω_p depends on the integration variables (x,y). To proceed further, it is convenient to take advantage of the distributive law for products and sums: exchanging the order of summation and products in Equation (5.7) is possible, provided that the index m is replaced by P indexes m_p (see Appendix A, Equation [A.5.2], with $a = u_p$ and $\xi = \Omega_p + \Phi_p$). Then, the product of P series is converted in P (nested) series whose terms are the product of P terms.

$$E_{pq}^{(s)}(\mathbf{r}) = -\frac{ik\exp(-ikr)}{4\pi r}E_pF_{pq}(\vartheta_i,\vartheta_s,\varphi_s)$$
$$\int_A \exp\{-i[\eta_x x + \eta_y y]\}\sum_{m_0=-\infty}^{+\infty}\cdots\sum_{m_{P-1}=-\infty}^{+\infty}$$
$$\left\{\exp\left[i\sum_{p=0}^{P-1} m_p[\Omega_p(x,y)+\Phi_p]\right]\prod_{p=0}^{P-1} J_{m_p}(u_p)\right\}dA. \tag{5.8}$$

Equation (5.8) shows that each generic index m_p is linked to the corresponding p-th tone of the WM function.

Now the integral in Equation (5.8) can be split into the sum of a P-infinity of elementary integrals:

$$E_{pq}^{(s)}(\mathbf{r}) = -\frac{ik\exp(-ikr)}{4\pi r}E_pF_{pq}(\vartheta_i,\vartheta_s,\varphi_s)$$
$$\sum_{m_0=-\infty}^{+\infty}\cdots\sum_{m_{P-1}=-\infty}^{+\infty}\left\{\left[\exp\left(i\sum_{p=0}^{P-1} m_p\Phi_p\right)\prod_{p=0}^{P-1} J_{m_p}(u_p)\right]\right.$$
$$\left.\int_A \exp\left[i\sum_{p=0}^{P-1} m_p\Omega_p(x,y)\right]\exp[-i(\eta_x x + \eta_y y)]\,dA'\right\}, \tag{5.9}$$

5.2. Analytic Derivation of the Scattered Field

each one amenable to an easy evaluation in closed form. For a scattering surface A with a rectangular projection over the mean plane $z = 0$, of dimensions X and Y along (x, y), respectively, it turns out that

$$E_{pq}^{(s)}(\mathbf{r}) = -\frac{ik\exp(-ikr)}{4\pi r} E_p F_{pq}(\vartheta_i, \vartheta_s, \varphi_s) XY$$

$$\sum_{m_0=-\infty}^{+\infty} \cdots \sum_{m_{P-1}=-\infty}^{+\infty} \left\{ \exp\left(i\sum_{p=0}^{P-1} m_p \Phi_p\right) \prod_{p=0}^{P-1} J_{m_p}(-\eta_z B C_p v^{-Hp}) \right.$$

$$\mathrm{sinc}\left[\left(-\eta_x + \kappa_0 \sum_{p=0}^{P-1} m_p v^p \cos\Psi_p\right)\frac{X}{2}\right]$$

$$\left.\mathrm{sinc}\left[\left(-\eta_y + \kappa_0 \sum_{p=0}^{P-1} m_p v^p \sin\Psi_p\right)\frac{Y}{2}\right]\right\}. \qquad (5.10)$$

Finally, it is convenient to highlight the scattered-field dependence on the WM function and geometry parameters by substituting the η values in Equation (5.10):

$$E_{pq}^{(s)}(\mathbf{r}) = -\frac{ik\exp(-ikr)}{4\pi r} E_p F_{pq}(\vartheta_i, \vartheta_s, \varphi_s) XY$$

$$\sum_{m_0=-\infty}^{+\infty} \cdots \sum_{m_{P-1}=-\infty}^{+\infty} \left\{ \exp\left(i\sum_{p=0}^{P-1} m_p \Phi_p\right) \right.$$

$$\prod_{p=0}^{P-1} J_{m_p}\left[k(\cos\vartheta_i + \cos\vartheta_s) B C_p v^{-Hp}\right]$$

$$\mathrm{sinc}\left[\left(-k(\sin\vartheta_i - \sin\vartheta_s\cos\varphi_s) + \kappa_0 \sum_{p=0}^{P-1} m_p v^p \cos\Psi_p\right)\frac{X}{2}\right]$$

$$\left.\mathrm{sinc}\left[\left(k\sin\vartheta_s\sin\varphi_s + \kappa_0 \sum_{p=0}^{P-1} m_p v^p \sin\Psi_p\right)\frac{Y}{2}\right]\right\}. \qquad (5.11)$$

The dependence on the electromagnetic parameters is present in the $F_{pq}(\cdot)$ function that includes the Fresnel reflection coefficients (see Chapter 4).

In the small-slope expansion, this function does not depend on the surface roughness (see Equations [4.51]) for the horizontal and vertical polarization, being evaluated with reference only to the surface mean plane.

Equation (5.11) states that the electromagnetic scattered field can be directly evaluated in terms of the WM and electromagnetic parameters as the superposition of modes, each one being characterized by P values of the P indexes m_p. Accordingly, we have a P-infinity of scattered modes: each radiated mode functionally appears to be generated from an appropriate combination of contribution from the P tones representing the surface roughness. The coefficients of the WM function characterize each mode: the Φ_ps contribute to each mode phase; the C_ps contribute to the amplitude term composed by P products of Bessel functions; the Ψ_ps contribute to the two sinc($\cdot$) terms.

The illumination footprint and electromagnetic incident wavenumber set the number of the tones, P, and the surface fundamental-tone wavenumber, κ_0 (see Chapter 3). These parameters, along with the value selected for the tones spacing, ν, fix in turn the surface spectrum region accounted for by the employed WM function, Equation (5.1). Then, according to the physical considerations reported in Chapter 3, these parameters fix the surface spectrum region, which is actually sensed by the incident electromagnetic field.

Finally, a comment on the higher-order statistics of the scattered field is due: Equation (5.11) turns out to be inappropriate for evaluation of the scattered-power density, because it does not lead to a (readable) closed-form expression of its mean-square value.

5.3. Scattered-Field Structure

The main importance of Equation (5.11) is to analytically represent the scattered field in terms of surface and sensor parameters. However, it appears too cumbersome to establish an easy link between the scattered field and the illumination and surface parameters. An exhaustive discussion is necessary to shed light on the implication of Equation (5.11). Accordingly, determination of the number of modes significantly contributing to the scattered field is in order, as well as evaluation of their amplitude, phase, and direction of propagation. As further insight, graphical results are presented for the simpler case $\psi_p = \varphi_p = 0$—that is, for a surface exhibiting a fractal

5.3.1. Number of Modes Significantly Contributing to the Scattered Field

Examination of the scattered modes described by Equation (5.11) is in order. Each mode is characterized by a string of P indexes, obtained by picking one term out of each series in Equation (5.11). For instance, in the case $P = 3$, a possible string is $[0, 1, 0]$, corresponding to the series terms characterized by $m_0 = 0, m_1 = 1, m_2 = 0$. The series terms composing the chosen mode are multiplied by each other, thus generating the overall amplitude and phase of the considered mode. A key role in any judgment on the mode amplitude is provided by the product of the Bessel functions appearing in Equation (5.11): its relevance is related to the Bessel function argument u_p and order m_p. In particular, $J_{m_p}(u_p) \cong 0$, as soon as $|u_p| < |m_p|$—that is, when the absolute value of the Bessel function index exceeds the absolute value of the Bessel function argument; this is in agreement with existing criteria devoted to determining the bandwidth of phase- or frequency-modulated signals. Hence, it is possible to set in Equation (5.11)

$$\prod_{p=0}^{P-1} J_{m_p}[k(\cos \vartheta_i + \cos \vartheta_s) BC_p v^{-Hp}] \to 0 \qquad (5.12)$$

whenever

$$\exists p : |m_p| > k(\cos \vartheta_i + \cos \vartheta_s)|BC_p|v^{-Hp}. \qquad (5.13)$$

Equation (5.12) implies that any mode whose string contains at least one index as in Equation (5.13), does not significantly contribute to the scattered field. Hence, the quantity $|u_p|$ can be read as the (adimensional) normalized roughness of the p-th tone of the surface: the higher the roughness of tones, the larger the number of modes that significantly contribute to the scattered field. This tone-normalized roughness depends on the surface parameters as well as on the illumination and observation directions and the electromagnetic-field wavelength (see Equation [5.13]): this confirms the widespread assumption that the surface roughness must be judged by also taking into account the adopted sensing system.

The normalized roughness $|u_p|$ of the p-th tone increases as the p-th tone amplitude, $BC_p v^{-Hp}$, increases as well. On average, the tone's normalized roughness decreases with p: the higher H, the more rapid this decrease, and fewer modes significantly contribute to the scattered field. Conversely, the lower H, the higher the tone roughness; this result confirms what was stated, from the geometric point of view, in Chapter 3.

The tone roughness increases as the frequency of the electromagnetic wave increases, thus confirming the key role played by the employed electromagnetic wavelength in any sensing instrument. In addition, the tone roughness increases as $(\cos\vartheta_i + \cos\vartheta_s)$ increases—that is, whenever the illumination and/or the observation directions approach the normal to the surface mean plane. Conversely, surfaces appear, or are remotely sensed as smoother, when observed at near-grazing angles. However, at near-grazing angles, the considered PO solution to the KA is not appropriate because surface shadowing is not taken into account.

The surface *overall* roughness is an important concept, intuitively related to the tones roughness. It is reasonable that a change of quoted parameters resulting in an increase in tones roughness will similarly increase the overall roughness. In addition, the higher the tones roughness, the larger the number of significant radiated modes. Accordingly, this number is the appropriate indicator of the surface overall roughness. Therefore, it is reasonable to identify the number of significant radiated modes with the surface overall normalized roughness.

Evaluation of the number of significant scattered modes M is in order. As already observed, for each index p, a negligible scattered field is obtained if $|u_p| < |m_p|$; thus, each m_p is practically upper bounded, and only $2\lceil |u_p| \rceil + 1$ choices are available for its value, where $\lceil \cdot \rceil$ indicates the upper integer ceiling. Accordingly, the total number of significant modes, M, is provided by the combination of the available values for each of the P indexes m_p, and is given by

$$M = \prod_{p=0}^{P-1} \left(2\lceil |u_p| \rceil + 1\right), \tag{5.14}$$

which clearly shows the dependence of the number of significant modes on the normalized surface roughness $|u_p|$.

Equation (5.14) provides a first conservative evaluation of the number of significant modes. However, the actual number is even lower, as presented hereafter.

5.3. Scattered-Field Structure

For $2|u_p| = 2k(\cos\vartheta_i + \cos\vartheta_s)|BC_p|v^{-Hp} \ll 1, \forall p$, Equation (5.14) shows that only the fundamental mode characterized by $m_p = 0, \forall p$ significantly contributes to the scattered field: this mode is individuated by the P-order string $[0,0,\ldots,0]$. In this case, Equation (5.11) can be simplified, and the scattered field is given by

$$E_{pq}^{(s)}(\mathbf{r}) = -\frac{ik\exp(-ikr)}{4\pi r} E_p F_{pq}(\vartheta_i, \vartheta_s, \varphi_s) XY$$

$$\prod_{p=0}^{P-1} J_0\left[k(\cos\vartheta_i + \cos\vartheta_s)BC_p v^{-Hp}\right]$$

$$\operatorname{sinc}\left[(-k(\sin\vartheta_i - \sin\vartheta_s \cos\varphi_s))\frac{X}{2}\right] \operatorname{sinc}\left[(k\sin\vartheta_s \sin\varphi_s)\frac{Y}{2}\right]. \tag{5.15}$$

In addition, for the assumed small values of the normalized roughness, $J_0(2k(\cos\vartheta_i + \cos\vartheta_s)BC_p v^{-Hp}) \cong 1$, $\forall p$, and Equation (5.15) further simplifies as

$$E_{pq}^{(s)}(\mathbf{r}) = -\frac{ik\exp(-ikr)}{4\pi r} E_p F_{pq}(\vartheta_i, \vartheta_s, \varphi_s) XY$$

$$\operatorname{sinc}\left[(-k(\sin\vartheta_i - \sin\vartheta_s \cos\varphi_s))\frac{X}{2}\right] \operatorname{sinc}\left[(k\sin\vartheta_s \sin\varphi_s)\frac{Y}{2}\right], \tag{5.16}$$

so that the scattered field is coincident with that appropriate to a flat surface. Accordingly, those fractal surfaces whose scattered field is dominated by the fundamental tone can be defined as *almost flat* (for that sensing instrument and geometry). For deterministic tone-amplitude coefficients and $H \neq 1$, the condition to assess the almost-flat behavior, $2k(\cos\vartheta_i + \cos\vartheta_s)|BC_p|v^{-Hp} \ll 1$, $\forall p$, simplifies because the tone amplitudes monotonically decrease with p, and reduces to enforce that $2k(\cos\vartheta_i + \cos\vartheta_s)|BC_0| \ll 1$.

For higher values of the normalized surface roughness, more tones significantly contribute to the scattered field; the first additional couple of modes to be included are characterized by $m_0 = \pm 1, m_p = 0$, $\forall p \neq 0$, so that they are individuated by the P-order strings $[1, 0, \ldots, 0]$ and $[-1, 0, \ldots, 0]$, respectively. Each mode within this couple exhibits a maximum-amplitude scattered field that is lower (on the average, for stochastic fractal surfaces) than that scattered by the specular mode, because $J_0[k(\cos\vartheta_i + \cos\vartheta_s)BC_0] > J_1[k(\cos\vartheta_i + \cos\vartheta_s)BC_1 v^{-H}]$ for low values of the Bessel function argument. For these new modes, Equation (5.11) simplifies, and the total scattered field is given by

$$E^{(s)}_{pq} = -\frac{ik \exp(-ikr)}{4\pi r} E_p F_{pq}(\vartheta_i, \vartheta_s, \varphi_s) XY$$

$$\left\{ \prod_{p=0}^{P-1} J_0\left[k(\cos\vartheta_i + \cos\vartheta_s)BC_p v^{-Hp}\right] \right.$$

$$\mathrm{sinc}\left[(-k(\sin\vartheta_i - \sin\vartheta_s \cos\varphi_s))\frac{X}{2}\right] \mathrm{sinc}\left[(k\sin\vartheta_s \sin\varphi_s)\frac{Y}{2}\right]$$

$$+ J_1[k(\cos\vartheta_i + \cos\vartheta_s)BC_0] \prod_{p=1}^{P-1} J_0[k(\cos\vartheta_i + \cos\vartheta_s)BC_p v^{-Hp}]$$

$$\left\{ \exp(i\Phi_0)\mathrm{sinc}\left[(-k(\sin\vartheta_i - \sin\vartheta_s \cos\varphi_s) + \kappa_0 \cos\Psi_p)\frac{X}{2}\right] \right.$$

$$\mathrm{sinc}\left[(k\sin\vartheta_s \sin\varphi_s + \kappa_0 \sin\Psi_p)\frac{Y}{2}\right]$$

$$- \exp(-i\Phi_0)\mathrm{sinc}\left[(-k(\sin\vartheta_i - \sin\vartheta_s \cos\varphi_s) - \kappa_0 \cos\Psi_p)\frac{X}{2}\right]$$

$$\left. \left. \mathrm{sinc}\left[(k\sin\vartheta_s \sin\varphi_s - \kappa_0 \sin\Psi_p)\frac{Y}{2}\right]\right\}\right\}. \tag{5.17}$$

For further roughness increases, the number of modes to be considered must be augmented by adding further mode couples.

For $2k(\cos\vartheta_i + \cos\vartheta_s)|BC_p|v^{-Hp} \gg 1, \forall p$, i.e., very rough surfaces, the number of significant modes M can be approximated as

$$M \cong \prod_{p=0}^{P-1} 2k(\cos\vartheta_i + \cos\vartheta_s)|BC_p|v^{-Hp}$$

$$\cong (2k(\cos\vartheta_i + \cos\vartheta_s)B)^P v^{-H\sum_{p=0}^{P-1} p} \tag{5.18}$$

$$= (2k(\cos\vartheta_i + \cos\vartheta_s)B)^P v^{-H\frac{P(P-1)}{2}},$$

wherein $\prod_{p=0}^{P-1}|C_p| \cong 1$ has been assumed, which is, on the average, reasonable because each factor C_p equals exactly 1 (deterministic case) or on the average (statistical case) (see Section 3.6). For large roughness and $\nu \to 1$ (fBm case, see Section 3.7), the number of significant modes, M, turns out to be approximately proportional to $[k(\cos\vartheta_i + \cos\vartheta_s)B]^P$ — that is, with a power law with the number of tones. Because M has been identified as the surface-roughness indicator, it turns out that the surface roughness dramatically increases with the number of tones.

5.3.2. Mode Directions of Propagation

The role of sinc($\cdot$) functions appearing in Equation (5.11) is in order. These functions provide (but for the $F_{pq}(\cdot)$ factor) the spatial distribution of the electromagnetic field for each mode: the smaller the sinc($\cdot$) function argument, the larger the radiated field. If the condition

$$\begin{cases} k\sin\vartheta_s \cos\varphi_s = k\sin\vartheta_i - \kappa_0 \sum_{p=0}^{P-1} m_p \nu^p \cos\Psi_p \\ k\sin\vartheta_s \sin\varphi_s = -\kappa_0 \sum_{p=0}^{P-1} m_p \nu^p \sin\Psi_p \end{cases} \quad (5.19)$$

is enforced, the sinc($\cdot$) functions attain their maximum (unitary) value. However, it is not granted that this value can be reached for (ϑ_s, φ_s) angles in the real space. Accordingly, modes may be divided into two categories: localized ones and diffused ones.

The localized modes are those whose argument of the sinc($\cdot$) functions may reach or be close to zero in the real (angular) region. This condition specifies a direction of propagation, as well as the main lobe of radiation.

On the other side, the diffused modes appear via the lateral lobes of the sinc($\cdot$) functions argument in the real (angular) region. The direction of propagation is difficult to individuate; in addition, the amplitude of the modes may be significantly lowered, so that the effective number of significant modes turns out to be smaller—or even much smaller— than the number provided by Equations (5.14). The main role in this decrease of the number of significant modes is played by the quantities $\kappa_0/k \left(\sum_{p=0}^{P-1} m_p \nu^p \cos\Psi_p \right)$, $\kappa_0/k \left(\sum_{p=0}^{P-1} m_p \nu^p \sin\Psi_p \right)$.

And examination of Equations (5.19) shows that an increase in the above quantities tends to move the corresponding mode in the side-lobe region of the sinc($\cdot$) function.

Direct expressions for the scattering angles ϑ_s, φ_s can be easily obtained by squaring and summing up Equations (5.19):

$$\begin{cases} \sin\vartheta_s = \sqrt{\left[\sin\vartheta_i - \frac{\kappa_0}{k}\sum_{p=0}^{P-1} m_p v^p \cos\Psi_p\right]^2 + \left(\frac{\kappa_0}{k}\right)^2 \left(\sum_{p=0}^{P-1} m_p v^p \sin\Psi_p\right)^2}, \\ \qquad\qquad\qquad\qquad\qquad\qquad\qquad\qquad\qquad\qquad \mathrm{Re}[\vartheta_s] \in [0, \pi/2] \\ \tan\varphi_s = \dfrac{\frac{\kappa_0}{k}\sum_{p=0}^{P-1} m_p v^p \sin\Psi_p}{\frac{\kappa_0}{k}\sum_{p=0}^{P-1} m_p v^p \cos\Psi_p - \sin\vartheta_i}, \quad \mathrm{Re}[\varphi_s] \in (-\pi, \pi]. \end{cases}$$

(5.20)

Inspection of Equations (5.20) shows possible appearance of complex values of ϑ_s.

It is noted that Equations (5.20) provide values of the couples (ϑ_s, φ_s) that are not necessarily solutions of Equations (5.19), due to the performed squaring operation. This ambiguity is resolved by verifying that each couple from Equations (5.20) is a solution of Equations (5.19).

For almost-flat surfaces ($m_p = 0$, $\forall p$), only one scattered mode is present, whose propagation direction turns out to be coincident with the specular one ($\vartheta_s = \vartheta_i$, $\varphi_s = 0$).

For increasing surface roughness, further couples of modes contribute significantly to the scattered field; the two modes belonging to the same couple do not generally propagate along symmetrical directions with respect to the specular mode: symmetry is recovered along ϑ only for normal incidence ($\vartheta_i = 0$), and along φ for y-directed surface tones ($\Psi_p = \pm\pi/2$).

As anticipated, numerical results become easier to read if the one-dimensional fractal surface along the x-axis and the field scattered in the plane $y = 0$ are considered. In the following discussion, this surface is modeled by a WM function with all Ψ_p equals, and two limiting cases are analyzed. In the first case, the surface is rough only along the x-axis, so that the surface roughness is aligned to the projection of the incident wave vector onto the scattering-surface mean plane. In the second case, the surface is rough only along the y-axis, so that the surface roughness is orthogonal to the projection of the incident wave vector onto the scattering-surface mean plane.

The first case is achieved by enforcing $\Psi_p = 0$ in Equation (5.20). The scattered modes attain their maximum along directions such that

$$\begin{cases} \sin \vartheta_s = \sin \vartheta_i - \dfrac{\kappa_0}{k} \sum_{p=0}^{P-1} m_p v^p, & \varphi_s = 0 \\ \sin \vartheta_s = -\left(\sin \vartheta_i - \dfrac{\kappa_0}{k} \sum_{p=0}^{P-1} m_p v^p \right), & \varphi_s = \pi \end{cases} \quad (5.21)$$

Note that Equations (5.21) can be obtained directly from Equations (5.19), so that no ambiguity on their solutions is present. Hence, all these modes propagate parallel to the plane $\varphi_s = 0, \pi$, this latter result being consistent with the Bragg theory. The fundamental tone is radiated along the specular direction ($\vartheta_s = \vartheta_i, \varphi_s = 0$), because the case $\varphi_s = \pi$ does not provide any solution to ϑ_s. Subsequent modes attain their maximum in the visible region if the condition

$$-(1 - \sin \vartheta_i) \leq \frac{\kappa_0}{k} \sum_{p=0}^{P-1} m_p v^p \leq (1 + \sin \vartheta_i) \quad (5.22)$$

is satisfied. The first couple, characterized by $m_0 = \pm 1, m_p = 0, \forall p \neq 0$, propagates along directions ϑ_s such that $\sin \vartheta_s = \sin \vartheta_i \mp \kappa_0/k$.

The second case is obtained by letting $\Psi_p = \pi/2$ in Equations (5.20). The scattered modes attain their maximum values along directions such that

$$\begin{cases} \sin \vartheta_s = \sqrt{(\sin \vartheta_i)^2 + \left(\dfrac{\kappa_0}{k} \sum_{p=0}^{P-1} m_p v^p \right)^2} \\ \sin \varphi_s = -\dfrac{\dfrac{\kappa_0}{k} \sum_{p=0}^{P-1} m_p v^p}{\sqrt{(\sin \vartheta_i)^2 + \left(\dfrac{\kappa_0}{k} \sum_{p=0}^{P-1} m_p v^p \right)^2}} \end{cases} \quad (5.23)$$

For $\vartheta_s = \vartheta_i, \varphi_s = 0$; moreover, the closer ϑ_s to ϑ_i, the closer to $\varphi_s = 0$ propagation occurs.

5.3.3. Modes Amplitude and Phase

The complex amplitude of each scattered mode depends on the product of the Bessel functions, the sinc(·) functions, and the phase term (see Equation [5.11]). Only those modes that significantly contribute to the scattered field (see Section 5.3.2) need to be considered. The larger the Bessel functions product, the more relevant is the mode amplitude. This implies that, at least for small roughness surfaces, modes with lower indexes are most significant. The role of the sinc($\cdot$) function has been presented (see Section 5.3.2) by individuating the two categories of localized and diffused modes.

The phase of the modes is determined by those of the corresponding WM surface expansion. A shift of any surface tone generates phase shift in all the scattered modes relevant to that tone, as well as a tilt of the corresponding radiation diagram: this is according to the FT relationship between aperture distribution and scattered fields.

5.4. Limits of Validity

Kirchhoff approximation holds if the surface mean radius of curvature is much greater than the electromagnetic incident wavelength. Small-slope approximation applies whenever the root-mean-square slope is much smaller than unity. However, the radius of curvature and the root-mean-square slope are not well defined for the mathematical WM surface presented in Chapter 3, which involves an infinite number of tones. For a physical WM surface, Equation (5.1), radius of curvature and the root-mean-square slope can be defined (see Chapter 3), and are related to H, B, ν, κ_0, P.

Relations found in Chapter 3 can be used to obtain a first approximate expression of the validity limits of the PO approach. In particular, substitution of the WM function in Equation (3.62) shows that the small-slope approximation applies whenever

$$\sigma'^2 = \kappa_0^2 \frac{B^2}{4} \frac{1 - \nu^{-2(H-1)P}}{1 - \nu^{-2(H-1)}} \ll 1. \tag{5.24}$$

A further connection is obtained by substituting the WM function in Equation (3.63) and requiring that the radius of curvature is much greater than the electromagnetic wavelength:

$$\frac{K^2}{k^2} \cong \frac{\sigma''^2}{k^2} = \frac{\kappa_0^4}{k^2} \frac{B^2}{8} \frac{1 - \nu^{-2(H-2)P}}{1 - \nu^{-2(H-2)}} \ll 1. \tag{5.25}$$

More-precise validity limits might be determined—for instance, by comparing theoretical results with numerical simulation of the scattered electromagnetic field from fractal surfaces.

5.5. Influence of Fractal and Electromagnetic Parameters over the Scattered Field

In this section, the influence of fractal parameters, κ_0, ν, P, B, H, on the scattered field is analyzed. For a systematic study, the WM coefficients are assumed to be not random: $C_p = 1, \Psi_p = 0, \Phi_p = 0$ for any p. The scattering surface is illuminated from the direction $\vartheta_i = \pi/4$, whereas the scattered field is displayed for any ϑ_s in the plane identified by $\varphi_s = 0, \pi$; this suggests assuming $-(\pi/2) < \vartheta_s \leq \pi/2$. The cut of the scattering surface with the plane $\varphi_s = 0, \pi$ is also shown in all subsequent figures.

The effect of the fractal parameters on the scattered field is first briefly described in a qualitative manner. Then, quantitative assessment via Equation (5.11) is graphically presented in Figures 5.2 through 5.7. The presence of the factor $F_{pq}(\cdot)$, which depends on the electromagnetic parameters in Equation (5.11), is omitted: the reason is that it also appears if classical surfaces are considered, it does not include any dependence on the fractal parameters, and it is not relevant to this discussion.

The reference scattering surface is characterized by the fractal parameters listed in Table 5.1. These parameters are typical for natural surfaces on the Earth when illuminated by a microwave instrument. Scattering surface and normalized scattered-power density are displayed in Figure 5.2, along with its scattered field. Then, in any subsequent subsection, only one of the surface fractal parameters is changed within a reasonable and significant range; its specific contribution to the scattered field is graphically displayed.

Plots in Figure 5.2 show that the maximum field scattered by the reference surface is attained in the specular direction. For scattering angles close to the specular direction, the scattered field approximately exhibits the sinc($\cdot$) behavior, thus showing that the specular mode dominates the scattering behavior. The scattered field is arranged in a series of lobes whose mean width is proportional to the ratio between the length of the illuminated area and the electromagnetic wavelength. The overall scattered field turns out to be obtained as superposition of the significant scattered modes, whose number and individual contribution cannot easily be individuated. Accordingly, with the considerations presented in Chapter 3 and Section 5.3.1, 4 tones are necessary to describe the surface, and 135 modes (localized or diffuse) have been considered in the numerical evaluation.

It is noted that each graph in Figures 5.2 through 5.7 refers to the particular ensemble element of the surface depicted in the same figure. This is convenient for many applications, but does not explicitly show the statistical

5 ◊ Scattering from WM Surfaces: PO Solution

TABLE 5.1 Parameters relevant to the illumination conditions and to the reference surface are considered in Figure 5.2 and are used to compare results in Figures 5.2 through 5.7.

Incidence angle (ϑ_i)	45°
Illuminated area (X, Y)	(1 m, 1 m)
Electromagnetic wavelength (λ)	0.1 cm
Hurst exponent (H)	0.8
Tone wave-number spacing coefficient (ν)	e
Overall amplitude-scaling factor (B)	0.01 m

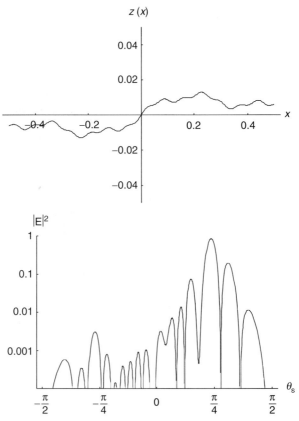

FIGURE 5.2 Plot of the profile, $y = 0$, of the reference surface holding the fractal parameters listed under Table 5.1 along with the graph of the corresponding normalized scattered power density in the plane $y = 0$.

5.5. Influence of Fractal and Electromagnetic Parameters

behavior of the ensemble itself. To get this additional information, as already pointed out in Section 1.6.1, the WM model parameters are changed according to their statistics, the associate scattered-density power for each element is computed, and the obtained values are properly processed. For instance, the expected value of the scattered-power density is obtained by averaging a sufficient number of graphs as that of Figure 5.2, each relative to one element of the surface ensemble. This is shown in Figure 5.8.

5.5.1. The Role of the Fundamental-Tone Wavenumber

As shown in Chapter 3, κ_0 is not an independent fractal parameter. Its value can be set according to the illuminated-surface dimension: as already shown

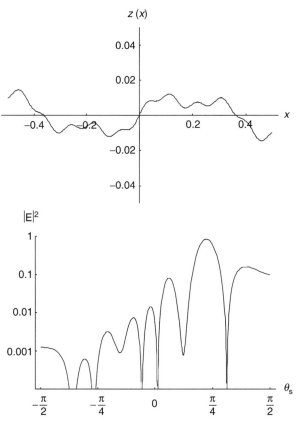

FIGURE 5.3a Modified versions of the reference surface are considered. As in Figure 5.2, surface cuts and corresponding normalized scattered power density graphs are depicted. Case $L_x = L_y = 0.5$ m.

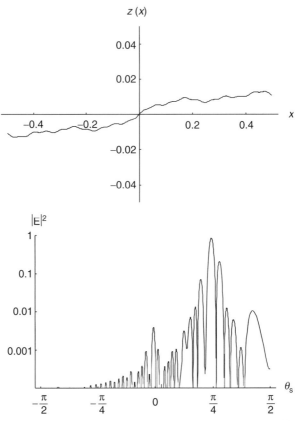

FIGURE 5.3b Modified versions of the reference surface are considered. As in Figure 5.2, surface cuts and corresponding normalized scattered power density graphs are depicted. Case $L_x = L_y = 2$ m.

(see Section 3.6), it is reasonable to set the fundamental-tone wavelength of the order of the maximum length of the surface, with an appropriate safety factor $\chi_1 \in (0, 1)$, the latter to be set equal to 0.1 in most critical cases. Accordingly:

$$\kappa_0 = \frac{2\pi}{\sqrt{X^2 + Y^2}} \chi_1. \qquad (5.26)$$

Then, instead of κ_0, from the electromagnetic-scattering viewpoint, it is more meaningful to consider the effect induced by a change of the

5.5. Influence of Fractal and Electromagnetic Parameters

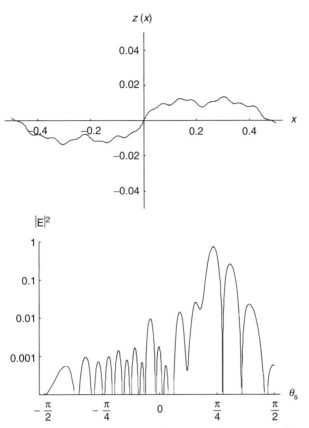

FIGURE 5.4a Modified versions of the reference surface are considered. As in Figure 5.2, surface cuts and corresponding normalized scattered power density graphs are depicted. Case $v = e - 0.5$.

scattering-surface dimensions. The scattered field evaluated by means of Equation (5.11) and relevant to different values of the surface dimensions is depicted in Figures 5.3.

Comparison of results in Figures 5.2 and 5.3 shows that the larger the illuminated area—that is, the smaller the value of κ_0—the narrower the radiated lobes; the scattered-power density oscillates more rapidly with ϑ_s. The side-lobes amplitude appears to be unaffected by κ_0. Equations (5.19) show that the smaller the value of κ_0, the higher the number of localized modes. In Figure 5.3a, 3 tones and 45 modes have been considered; in Figure 5.3b, 5 tones and 405 modes have been considered.

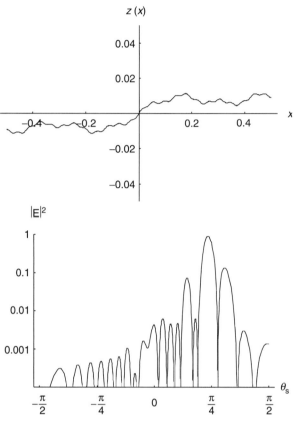

FIGURE 5.4b Modified versions of the reference surface are considered. As in Figure 5.2, surface cuts and corresponding normalized scattered power density graphs are depicted. Case $v = e + 0.5$.

5.5.2. The Role of the Tone Wave-Number Spacing Coefficient

As shown in Section 3.6, v is the fractal parameter that controls the tone wave-number spacing. Its value sets the degrees of similarity between the considered WM function and an appropriate fBm process of appropriate fractal parameters. Its value can be set to any irrational number >1.

The scattered field evaluated by means of Equation (5.11) and relevant to different values of v is reported in Figures 5.4.

Comparison of Figures 5.2 and 5.4 shows that the tone spacing marginally influences the scattered field. A smoother overall shape for the

5.5. Influence of Fractal and Electromagnetic Parameters

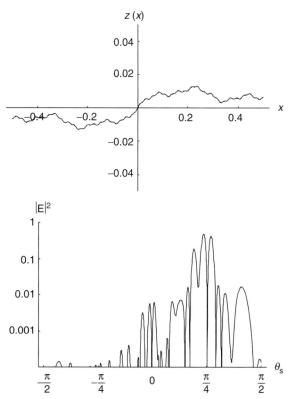

FIGURE 5.5a Modified versions of the reference surface are considered. As in Figure 5.2, surface cuts and corresponding normalized scattered power density graphs are depicted. Case $\lambda = 0.05$ m.

scattered-power density is obtained for lower ν. However, the ν value can greatly influence the number of tones that are required to represent the surface: thus, the number of significant modes, and consequently the computational time for the scattered field, is greatly influenced by the choice of ν. In Figure 5.4a, 5 tones and 405 modes have been considered; in Figure 5.4b, 4 tones and 135 modes have been considered. In conclusion, it is confirmed that ν is a fractal parameter that can be set to obtain reliable discrete approximation of the continuous fBm spectral behavior, whereas its value could not be set too close to 1 to allow efficient evaluation of the scattered field.

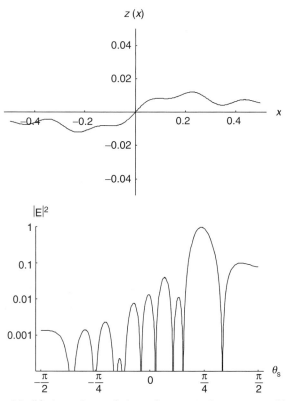

FIGURE 5.5b Modified versions of the reference surface are considered. As in Figure 5.2, surface cuts and corresponding normalized scattered power density graphs are depicted. Case $\lambda = 0.2$ m.

5.5.3. The Role of the Number of Tones

As shown in Chapter 3, P is not an independent fractal parameter. Its value can be set according to the incident wavelength and illuminated-surface dimension: as already shown (see Section 3.6), it is reasonable to set the fundamental-tone wavelength of the order of the electromagnetic wavelength, considering a further appropriate safety factor $\chi_2 \in (0,1)$, the latter is usually set to 0.1. Accordingly,

$$P = \left\lceil \frac{\ln\left(\sqrt{X^2 + Y^2}/\chi_1\chi_2\lambda\right)}{\ln \nu} \right\rceil + 1. \tag{5.27}$$

5.5. Influence of Fractal and Electromagnetic Parameters

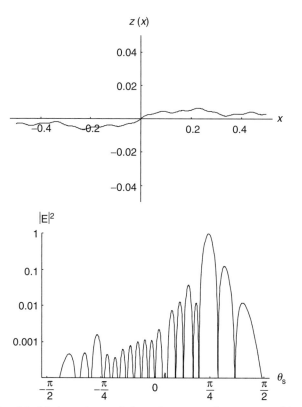

FIGURE 5.6a Modified versions of the reference surface are considered. As in Figure 5.2, surface cuts and corresponding normalized scattered power density graphs are depicted. Case $B = 0.005$ m.

The number of tones increases logarithmically with the ratio between the surface dimension to the electromagnetic wavelength. The lower the safety factors or tone spacing, the higher the number of tones.

In view of Equation (5.27) and the results of Section 5.3.3, instead of P it is more meaningful to consider changes induced by varying the electromagnetic wavelength λ. The scattered field evaluated by means of Equation (5.11) and relevant to different values of λ is reported in Figures 5.5.

Comparison of Figures 5.2 and 5.5 shows that the larger the number of tones—that is, the smaller the electromagnetic wavelength—the narrower the radiated lobes. For smaller λ, a finer description of the surface

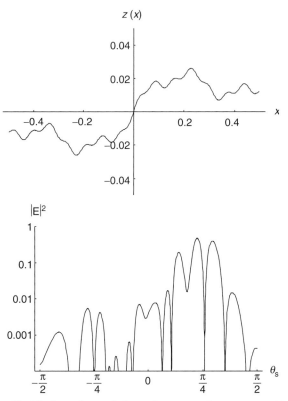

FIGURE 5.6b Modified versions of the reference surface are considered. As in Figure 5.2, surface cuts and corresponding normalized scattered power density graphs are depicted. Case $B = 0.02$ m.

is required, including more tones; consequently, the number of modes increases. In Figure 5.5a, 5 tones and 567 modes have been considered; in Figure 5.5b, 3 tones and 27 modes have been considered.

5.5.4. *The Role of the Overall Amplitude-Scaling Factor*

The overall amplitude-scaling factor B directly influences the surface roughness, Equations (5.1). Normalized tone roughness turns out to be proportional to B. Hence, the number of significant modes, M, increases with B. Conversely, B does not affect the number of significant radiated modes, the latter depending only on mode directions of propagation.

5.5. Influence of Fractal and Electromagnetic Parameters

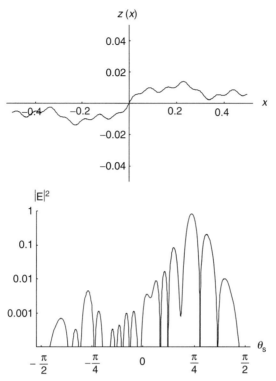

FIGURE 5.7a Modified versions of the reference surface are considered. As in Figure 5.2, surface cuts and corresponding normalized scattered power density graphs are depicted. Case $H = 0.7$ m.

In Figures 5.6, some graphs are reported of the scattered field relevant to different values of the overall amplitude-scaling factor B.

Comparison of Figures 5.2 and 5.6 shows that the larger the overall amplitude-scaling factor B, the larger the number of modes that significantly contribute to the scattered field. Accordingly, the scattered field resembles the sinc($\cdot$) behavior for lower B; conversely, for higher B, the sinc($\cdot$) behavior is lost. Note that the number of modes may increase, also varying other fractal parameters; but, accordingly with the considerations presented in Section 5.3, the disappearance of the sinc($\cdot$) behavior is more evident if B is increased. In Figure 5.6a, 4 tones and 81 modes have been considered; in Figure 5.6b, 4 tones and 189 modes have been considered.

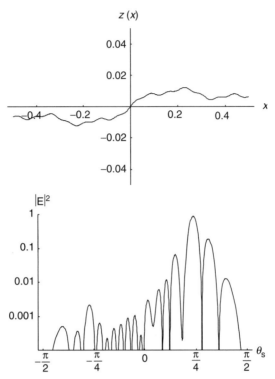

FIGURE 5.7b Modified versions of the reference surface are considered. As in Figure 5.2, surface cuts and corresponding normalized scattered power density graphs are depicted. Case $H = 0.9$ m.

5.5.5. The Role of the Hurst Exponent

As shown in Chapter 3, the Hurst coefficient is related to the fractal dimension. The lower H, the higher the fractal dimension, and the higher the number of significant modes that enter into the evaluation of the scattered field.

In Figures 5.7, two graphs are reported, relevant to surfaces with higher, $H = 0.9$, and lower, $H = 0.7$, values of the Hurst coefficient with respect to the intermediate one $H = 0.8$, reported in Figure 5.2 and relative to the reference surface.

Comparison of Figures 5.2 and 5.7 shows that the smaller the Hurst coefficient, the higher the lateral lobes. In both Figures 5.7a and 5.7b, 4 tones and 135 modes have been considered.

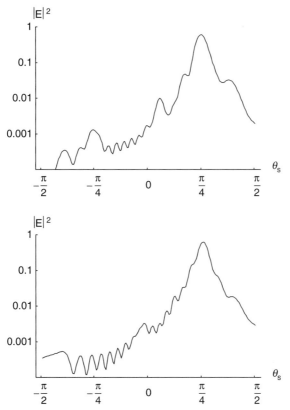

FIGURE 5.8 Plots of the normalized scattered power density relevant to a set of surfaces which have been obtained by randomizing the surface coefficients of the reference surface. Thirty-two realizations have been considered. top $\nu = e$; bottom $\nu = e - 0.5$.

5.6. Statistics of the Scattered Field

The scattered field evaluated in this chapter is amenable to being used within any study concerning random natural surfaces. Within this framework, it can be used to evaluate the scattered-field statistics starting from the generation of a set of surfaces with prescribed fractal parameters. For each surface of the set, the scattered field can be evaluated in terms of amplitude and phase. Then, for each position, the statistics of the scattered field can be obtained by appropriate processing of the computed scattered field. Note that this procedure does not prescribe any statistics to the scattered field: these come out by examination of the scattered field. Obviously, obtained results are confined by the fractal assumption of the surface. The scattered-power density relevant to a set of WM surfaces obtained by randomizing the WM coefficients of the reference surface is shown in Figure 5.8. A smoother power-density diagram is obtained if the number of modes is increased by diminishing the tone wave-number spacing coefficient ν from e to $e - 0.5$.

5.7. References and Further Readings

The physical-optics solution to the scattering problem relevant to classical surfaces is derived in detail in several books, including studies by Tsang, Kong, and Shin (1985) and by Ulaby and Dobson (1989). Details on the WM function are given in works by Falconer (1990), Feder (1988), and Mandelbrot (1983). Several graphs devoted to presenting the WM profiles and their characteristics are in a book by Berry and Lewis (1980). Scattering from WM profiles is studied in a work by Jaggard (1990). The list of papers written by the authors of this book and concerning the subject of this chapter is found in Appendix C.

CHAPTER 6

Scattering from Fractional Brownian Surfaces: Physical-Optics Solution

6.1. Introduction and Chapter Outline

In this chapter, a method is presented that allows evaluation of the electromagnetic (expected) power density scattered from natural rough surfaces described by means of fractional Brownian motion (fBm) processes. The Physical-Optics (PO) solution under the Kirchhoff Approximation (KA) to the diffraction integral is explored.

The natural surface is modeled by means of the fBm process and illuminated by a plane wave; in Section 6.2 the mean-square value of the scattered far field is analytically evaluated in a closed form within the KA in the small-slope regime. In Section 6.3 the final result is arranged in two different expressions that can be chosen according to the values of the surface fractal parameters. These expressions simplify in some special cases that are discussed in detail in Section 6.4. The backscattering coefficient—that is, the normalized *radar cross section* (RCS) of the surface—is also provided in Section 6.5. A theoretical discussion on the validity of this model is carried out in Section 6.6. Dependence of the scattered power density on the fBm and electromagnetic parameters is presented in Section 6.7. Key references and suggestions for further readings are reported in Section 6.8.

6 ◊ Scattering from Fractional Brownian Surfaces: Physical-Optics Solution

The fBm surface model is employed not really for its mathematical convenience, but rather because it best suits natural rough surfaces, according to what was reported in Chapter 3. As a matter of fact, the power-law spectrum of the fBm process has an exponent α (see Equation [3.27]), which is limited to the interval $2 < \alpha < 4$, as can be verified by recalling that $0 < H < 1$. This is also consistent with all measured soil-surface spectra, which exhibit power laws with $\alpha \approx 3$. The benefit that the fBm fractal characterization allows the analytic evaluation of the scattered field is also appealing: note that other generic power-law descriptions require aid of numerical methods or Monte Carlo simulations.

6.2. Scattered Power-Density Evaluation

In this section, a closed-form solution is derived for the expected value of the electromagnetic power density scattered from a natural surface modeled by means of an fBm random process: the solution is obtained according to the KA in the small-slope regime introduced in Section 4.6. Geometry of the scattering problem is shown in Figure 6.1.

The starting point is the formulation of the scattered field expressed in terms of Equation (4.47); then the mean-square value of the scattered field

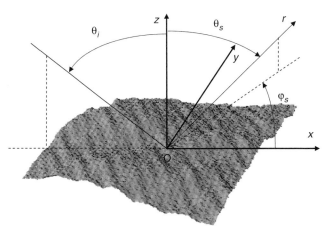

FIGURE 6.1 Geometry of the scattering problem. Cartesian and polar reference systems relevant to the scattering surface are depicted.

6.2. Scattered Power-Density Evaluation

is evaluated as

$$\langle |E_{pq}|^2 \rangle = \frac{k^2 |E_p|^2 |F_{pq}|^2}{(4\pi r)^2}$$

$$\times \int_{-\frac{X}{2}}^{\frac{X}{2}} \int_{-\frac{X}{2}}^{\frac{X}{2}} \int_{-\frac{Y}{2}}^{\frac{Y}{2}} \int_{-\frac{Y}{2}}^{\frac{Y}{2}} \exp\left[-i\eta_x(x-x') - i\eta_y(y-y')\right]$$

$$\times \langle \exp\left[-i\eta_z(z-z')\right]\rangle dx dx' dy dy' \quad (6.1)$$

where (x, y, z) and (x, y', z') are two generic points belonging to the surface. The projection of the area onto the mean plane is assumed to have a rectangular shape of dimensions X and Y; the factor $\boldsymbol{\eta} = (\eta_x, \eta_y, \eta_z)$ depends on the illumination and observation directions and is expressed in the Cartesian (x, y, z) reference system as

$$\boldsymbol{\eta} = \mathbf{k}_i - \mathbf{k}_s = \hat{\mathbf{x}} k (\sin \vartheta_i - \sin \vartheta_s \cos \varphi_s) - \hat{\mathbf{y}} k \sin \vartheta_s \sin \varphi_s$$
$$- \hat{\mathbf{z}} k (\cos \vartheta_i + \cos \vartheta_s) \quad (6.2)$$

(see Chapter 4).

The expected value appearing inside the integral, $\langle \exp\left[-i\eta_z(z-z')\right]\rangle$, is the characteristic function of the process of increments, $z - z'$: it has been shown in Chapter 3 that an fBm fractal model leads to a stationary and Gaussian increment process. Accordingly, the expected value in Equation (6.1) can be evaluated in terms of the structure function $Q(\tau)$, as shown in Chapter 2. For isotropic surfaces,

$$\langle \exp\left[-i\eta_z(z-z')\right]\rangle = \exp\left[-\frac{1}{2}\eta_z^2 Q(\tau)\right]. \quad (6.3)$$

Substituting Equation (6.2) in Equation (6.1), enforcing the coordinates transformation

$$\begin{cases} x - x' = \tau \cos \varphi \\ y - y' = \tau \sin \varphi \\ x = x \\ y = y \end{cases}, \quad (6.4)$$

and integrating with respect to x and y, Equation (6.1) leads to

$$\langle |E_{pq}|^2 \rangle = \frac{k^2 |E_p|^2 |F_{pq}|^2}{(4\pi r)^2} A \iint_A \exp\left[-i\tau \left(\eta_x \cos \varphi + \eta_y \sin \varphi\right)\right]$$

$$\exp\left[-\frac{1}{2}\eta_z^2 Q(\tau)\right] \tau \, d\varphi \, d\tau, \quad (6.5)$$

where A is the projection of the area of the illuminated surface onto the mean plane.

In Equation 6.5 the integration over φ can be performed according to Equation (A.6.6) with $a = \tau\eta_x$ and $b = \tau\eta_y$. In order to evaluate the remaining integration with respect to τ, the expression for the surface-structure function $Q(\tau) = s^2\tau^{2H}$ appropriate to the isotropic fBm process (see Equation [3.13]) must be used. Examination of Equation (6.5) allows extending the integration limits to the whole plane, provided that the linear size l of the illuminated area is such that

$$\eta_z^2 Q(l) \gg 1. \tag{6.6}$$

Accordingly:

$$\langle |E_{pq}|^2 \rangle = \frac{k^2 \left|E_p^{(i)}\right|^2 |F_{pq}|^2}{(4\pi r)^2} 2\pi A \int_0^\infty J_0(\eta_{xy}\tau) \exp\left[-\frac{1}{2}\eta_z^2 Q(\tau)\right] \tau\, d\tau, \tag{6.7}$$

where

$$\eta_{xy} = \sqrt{\eta_x^2 + \eta_y^2}. \tag{6.8}$$

In passim, it is noted that Equation (6.7) is also valid if classical-surface models are employed. However, for classical surfaces, the surface variance is finite, $Q(\tau)$ is limited, and it does not always exist a surface linear size l that verifies Equation (6.6). Moreover, use of a classical model employing a stationary stochastic Gaussian process for surfaces of infinite extent leads to

$$\langle |E_{pq}|^2 \rangle = \frac{k^2 \left|E_p^{(i)}\right|^2 |F_{pq}|^2}{(4\pi r)^2} \exp\left(-\eta_z^2 \sigma^2\right)$$

$$A\left[\delta(\eta_x)\delta(\eta_y) + \sum_{n=1}^\infty \frac{\eta_z^{2n}}{n!} W^{(n)}(\eta_{xy})\right]. \tag{6.9}$$

A closed-form expression for scattered power requires a similarly closed-form expression for the generalized power spectra $W^{(n)}(\kappa)$ in Equation (6.9). Accordingly, with the results shown in Chapter 2, this condition is met only in few specific cases (see Table 2.4); even in these cases, the corresponding expressions often include nonelementary functions. More important, as already noted, these models are usually not appropriate for natural surfaces, fractal geometry definitively being a better candidate.

6.2. Scattered Power-Density Evaluation

Substitution of the expression for $Q(\tau)$, Equation (3.13) in Equations (6.6) and (6.7), leads to

$$\eta_z^2 s^2 l^{2H} = k^2 (\cos\vartheta_i + \cos\vartheta_s)^2 s^2 l^{2H} \gg 1, \qquad (6.10)$$

and

$$\left\langle |E_{pq}|^2 \right\rangle = \frac{k^2 |E_p|^2 |F_{pq}|^2}{(4\pi r)^2} 2\pi A \int_0^\infty J_0\left(\eta_{xy}\tau\right) \exp\left[-\frac{1}{2}\eta_z^2 s^2 \tau^{2H}\right] \tau d\tau. \qquad (6.11)$$

Hence, for any sufficiently large fractal surface, Condition (6.10) is fulfilled, and Equation (6.11) holds.

As a last remark, note that the integral of Equation (6.11), initially space limited, now spans on all the τ interval from 0 to ∞ due to the presence of the exponential decaying factor. Accordingly, a mathematical fractal can be used, because computation of the scattered field automatically sets a limit on the relevant large distances ranges, i.e., on the low frequency fractal content (infrared catastrophe, see Chapter 3).

Solution of the integral in Equation (6.11) deserves an exhaustive discussion. In fact, a closed-form solution can be obtained only by using appropriate series expansions for the integrand function. Two options are available, and are detailed in the following discussion: either the Bessel or the exponential function is expanded in power series. Different power-series expansion obviously calls for different convergence rationales, thus implying that different closed-form solutions must be used according to the values of parameters representing the natural surface.

6.2.1. Persistent fBm

Power-series expansion of the zero-order Bessel function, according to Equation (A.6.1), with $t \equiv \eta_{xy}\tau$ gives

$$\int_0^\infty J_0\left(\eta_{xy}\tau\right) \exp\left[-\frac{1}{2}\eta_z^2 s^2 \tau^{2H}\right] \tau d\tau$$

$$= \int_0^\infty \sum_{n=0}^\infty \frac{(-1)^n \eta_{xy}^{2n}}{2^{2n}(n!)^2} \tau^{2n} \exp\left[-\frac{1}{2}\eta_z^2 s^2 \tau^{2H}\right] \tau d\tau. \qquad (6.12)$$

Integration of the series term by term, and use of Equation (A.6.2) with $u = \eta_z^2 s^2/2$, $v = 2H$ and $w = 2n + 1$, leads to

$$\int_0^\infty J_0(\eta_{xy}\tau) \exp\left[-\frac{1}{2}\eta_z^2 s^2 \tau^{2H}\right] \tau d\tau$$

$$= \frac{1}{2H} \sum_{n=0}^\infty \frac{(-1)^n}{2^{2n}(n!)^2} \cdot \Gamma\left(\frac{n+1}{H}\right) \frac{\eta_{xy}^{2n}}{\left(\frac{\sqrt{2}}{2}|\eta_z|s\right)^{\frac{2n+2}{H}}}. \quad (6.13)$$

Note that applicability of integration of the series term by term is not ensured in this case, because integration in Equation (6.11) spans over an unbounded interval. Accordingly, convergence of the series in Equations (6.12) and (6.13) must be checked.

It turns out that this series is convergent only for $H \geq 1/2$—that is, for persistent (see Section 3.5) fBm processes. However, for any H, Equation (6.12) can still be used under appropriate circumstances. In fact, the presence of the exponential term in the integrand allows us to limit the integration interval to values of τ such that $\eta_z s \tau^H < 3$—that is, for

$$\tau < \frac{3}{(\eta_z s)^{1/H}} = \frac{3}{[k(\cos\vartheta_i + \cos\vartheta_s)s]^{1/H}}. \quad (6.14)$$

Then, if

$$\frac{\eta_{xy}^H}{|\eta_z|s} = \frac{\left(\sin^2\vartheta_i + \sin^2\vartheta_s - 2\sin\vartheta_i \sin\vartheta_s \cos\varphi_s\right)^{H/2}}{k^{1-H}|(\cos\vartheta_i + \cos\vartheta_s)|s} \ll 1, \quad (6.15)$$

it turns out that the Bessel function argument is much smaller than unity, so that the series expansion of the Bessel function in Equation (6.11) can be truncated after a few terms. Hence, if Condition (6.14) is verified, Equation (6.13), truncated after a few terms, can be used irrespective of the value of H.

6.2.2. Antipersistent fBm

The zero-order Bessel function, $J_0(\cdot)$, is the real part of the zero-order Hankel function of the first kind, $H_0^{(1)}(\cdot)$; the integral appearing in

6.2. Scattered Power-Density Evaluation

Equation (6.11) can be written as

$$\int_0^\infty J_0(\eta_{xy}\tau)\exp\left[-\frac{1}{2}\eta_z^2 s^2\tau^{2H}\right]\tau d\tau$$

$$= \text{Re}\left\{\int_0^\infty H_0^{(1)}(\eta_{xy}\tau)\exp\left(-\frac{1}{2}\eta_z^2 s^2\tau^{2H}\right)\tau d\tau\right\}. \quad (6.16)$$

It is convenient to modify the integral inside the Real Part. This is accomplished by analytical continuation of t in the complex plane $p = t + i.$ and evaluation of the integral

$$\oint_C H_0^{(1)}(\eta_{xy}p)\exp\left(-\frac{1}{2}\eta_z^2 s^2 p^{2H}\right)p\,dp$$

$$= \int_0^\infty H_0^{(1)}(\eta_{xy}\tau)\exp\left(-\frac{1}{2}\eta_z^2 s^2\tau^{2H}\right)\tau d\tau$$

$$- \int_0^\infty H_0^{(1)}(\eta_{xy}i\xi)\exp\left[-\frac{1}{2}\eta_z^2 s^2 (i\xi)^{2H}\right]i\xi\,d(i\xi) = 0 \quad (6.17)$$

where the integration contour is depicted in Figure 6.2. The value of the integral is equal to zero because the ingtegrand exhibits no singularities

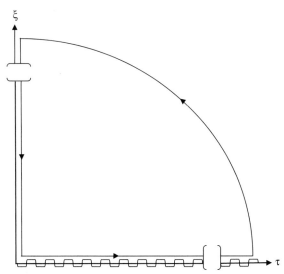

FIGURE 6.2 Integration closed contour in the complex plane for the integral in Eq. (6.17). A branch cut along the whole real positive axis is enforced to render the function p^ν single valued.

6 ◊ Scattering from Fractional Brownian Surfaces: Physical-Optics Solution

inside the integration contour, and the only contribution of the circle at infinity is zero. Use of Equation (6.17) allows the desired modofication, with the additional use of Equation (A.6.3) with $x \equiv \eta_{xy}\xi$, reaching the final result:

$$\int_0^\infty J_0(\eta_{xy}\tau) \exp\left[-\frac{1}{2}\eta_z^2 s^2 \tau^{2H}\right] \tau d\tau$$

$$= \text{Re}\left\{-\frac{2}{\pi i}\int_0^\infty K_0(\eta_{xy}\xi) \exp\left[-\frac{1}{2}\eta_z^2(i\xi)^{2H}\right] \xi d\xi\right\}. \quad (6.18)$$

Because the zero-order Kelvin function, $K_0(\cdot)$, decays exponentially for large values of its argument, expansion of the complex exponential in the second member of Equation (6.18) is allowed:

$$\int_0^\infty J_0(\eta_{xy}\tau) \exp\left[-\frac{1}{2}\eta_z^2 s^2 \tau^{2H}\right] \tau d\tau$$

$$= \text{Re}\left\{\frac{2}{\pi}\int_0^\infty K_0(\eta_{xy}\xi) \sum_{n=0}^\infty \frac{(-1)^{n+1}}{n!} i^{2Hn-1} \left(\frac{1}{2}\eta_z^2 s^2\right)^n \xi^{2Hn+1} d\xi\right\}. \quad (6.19)$$

The series is integrated term by term, and Equation (A.6.4) with $b = \eta_{xy}$, and $\mu = 2Hn + 1$ is used to obtain

$$\int_0^\infty J_0(\eta_{xy}\tau) \exp\left[-\frac{1}{2}\eta_z^2 s^2 \tau^{2H}\right] \tau d\tau$$

$$= \text{Re}\left\{\frac{2}{\pi}\sum_{n=0}^\infty \frac{(-1)^{n+1}}{n!} i^{2Hn+1} \left(\frac{1}{2}\eta_z^2 s^2\right)^n 2^{2Hn} b^{-2Hn-2} \Gamma^2(1+Hn)\right\}. \quad (6.20)$$

Taking the real part and use of Equation (A.6.5) leads to

$$\int_0^\infty J_0(\eta_{xy}\tau) \exp\left[-\frac{1}{2}\eta_z^2 s^2 \tau^{2H}\right] \tau d\tau$$

$$= 2H\sum_{n=1}^\infty \frac{(-1)^{n+1} 2^{2nH}}{n!} \frac{n\Gamma(1+nH)}{\Gamma(1-nH)} \frac{(\frac{\sqrt{2}}{2}|\eta_z|s)^{2n}}{\eta_{xy}^{2nH+2}}. \quad (6.21)$$

Note that also in this case applicability of integration by series is not ensured, because integration spans over an unbounded interval. Accordingly, convergence of the series in Equation (6.21) must be checked.

It turns out that Equation (6.21) is convergent only for $0 < H < 1/2$—that is, for antipersistent (see Section 3.5) fBm processes. However, for any H, the equation can still be used under appropriate circumstances. In fact, the presence of the Kelvin function in the integrand allows us to limit the integration interval to values of τ such that in the integrand of Equation (6.18), $\eta_{xy}\tau < 3$—that is, for

$$\tau < \frac{3}{\eta_{xy}} = \frac{3}{k\left(\sin^2 \vartheta_i + \sin^2 \vartheta_s - 2\sin \vartheta_i \sin \vartheta_s \cos \varphi_s\right)^{1/2}}. \quad (6.22)$$

Then, if

$$\frac{|\eta_z|s}{\eta_{xy}^H} = \frac{k^{1-H}|(\cos \vartheta_i + \cos \vartheta_s)|s}{\left(\sin^2 \vartheta_i + \sin^2 \vartheta_s - 2\sin \vartheta_i \sin \vartheta_s \cos \varphi_s\right)^{H/2}} \ll 1, \quad (6.23)$$

it turns out that the exponential-function argument is much smaller than unity, so that the series expansion of the exponential function in the second member of Equation (6.18) can be truncated after a few terms. Hence, if Condition (6.23) is verified, Equation (6.21), truncated after a few terms, can be used, irrespective of the H value.

6.3. Scattered Power Density

Results from the previous section are expressed here in terms of physical parameters relevant to the scattering problem; in particular, the scattering angular behavior is emphasized. The rationale to use obtained formulas for the scattered-power evaluation is presented.

If the Hurst coefficient, H, and the topothesy, T, are the parameters used to describe the fractal surface, and if Equation (4.53) is considered, the scattered-power density can be evaluated by substituting Equation (3.5) in

Equations (6.13) and (6.21), thus getting

$$\langle |E_{pq}|^2 \rangle = \frac{|E_p|^2 |F_{pq}|^2}{(4\pi r)^2} 2\pi A \frac{k^2 T^2}{2H} \sum_{n=0}^{\infty} \frac{(-1)^n}{2^{2n}(n!)^2} \Gamma\left(\frac{n+1}{H}\right)$$

$$\frac{(\sin^2 \vartheta_i + \sin^2 \vartheta_s - 2\sin \vartheta_i \sin \vartheta_s \cos \varphi_s)^n (kT)^{2n}}{[\frac{\sqrt{2}}{2}|(\cos \vartheta_i + \cos \vartheta_s)|kT]^{\frac{2n+2}{H}}}, \quad (6.24)$$

$$\langle |E_{pq}|^2 \rangle = \frac{|E_p|^2 |F_{pq}|^2}{(4\pi r)^2} 2\pi A 2 H k^2 T^2 \sum_{n=1}^{\infty} \frac{(-1)^{n+1} 2^{2nH}}{(n-1)!} \frac{\Gamma(1+nH)}{\Gamma(1-nH)}$$

$$\frac{(\frac{\sqrt{2}}{2}|(\cos \vartheta_i + \cos \vartheta_s)|kT)^{2n}}{\left[(\sin^2 \vartheta_i + \sin^2 \vartheta_s - 2\sin \vartheta_i \sin \vartheta_s \cos \varphi_s)(kT)^2\right]^{nH+1}} \quad (6.25)$$

for persistent and antipersistent fBm, respectively.

These two expressions are, to some extent, complementary. The series in Equation (6.24) is convergent only for $H \geq 1/2$; however, it is an asymptotic expansion of Integral (6.11)

$$\frac{\left[(\sin^2 \vartheta_i + \sin^2 \vartheta_s - 2\sin \vartheta_i \sin \vartheta_s \cos \varphi_s)^{1/2} kT\right]^H}{|(\cos \vartheta_i + \cos \vartheta_s)|kT} \to 0, \quad (6.26)$$

irrespective of the value of H. Conversely, the series in Equation (6.25) is convergent only for $0 < H \leq 1/2$; however, it is an asymptotic expansion of Integral (6.11) for

$$\frac{\left[(\sin^2 \vartheta_i + \sin^2 \vartheta_s - 2\sin \vartheta_i \sin \vartheta_s \cos \varphi_s)^{1/2} kT\right]^H}{|(\cos \vartheta_i + \cos \vartheta_s)|kT} \to \infty, \quad (6.27)$$

irrespective of the value of H. Furthermore, the convergence rate of Series (6.24) increases as the factor in Equation (6.26) decreases, whereas the convergence rate of Series (6.25) increases as the factor in Equation (6.11) increases.

6.4. Scattered Power Density: Special Cases

Equations (6.24) and (6.25) simplify in some special cases: scattering in the specular direction ($\eta_{xy} = 0$), Brownian surfaces ($H = 1/2$), marginally fractal surfaces ($H \to 1$), and quasi-smooth surfaces ($kT \ll 1$). One single expression can be used to evaluate the scattered-power density in each of these special cases, and these expressions can sometimes be derived in a more straightforward way with respect to Equations (6.24) and (6.25). Results for these special cases are discussed in the following sections.

6.4.1. Scattering in the Specular Direction

In the specular direction:

$$\begin{cases} \vartheta_i = \vartheta_s \\ \varphi_s = 0 \end{cases}, \tag{6.28}$$

so that $\eta_x = \eta_y = \eta_{xy} = 0$ and $\eta_z = 2k\cos\vartheta_i$: in this case, Equation (6.26) is always verified and use of Equation (6.24) is appropriate, thus getting

$$\langle |E_{pq}|^2 \rangle = \frac{|E_p|^2 |F_{pq}|^2}{(4\pi r)^2} 2\pi A \frac{k^2 T^2}{2H} \frac{\Gamma\left(\frac{1}{H}\right)}{(\sqrt{2}kT\cos\vartheta_i)^{2/H}}. \tag{6.29}$$

6.4.2. Brownian Surfaces (H=1/2)

By definition, a Brownian surface is obtained by setting $H = 1/2$. Its intersection with an arbitrary vertical plane is the graph of a Wiener process.

Starting directly from Equation (6.11), specified for $H = 1/2$, and Equation (A.6.7) with $b = \eta_{xy}, u = \eta_z^2 s^2/2$, the result

$$\langle |E_{pq}|^2 \rangle = \frac{|E_p|^2 |F_{pq}|^2 k^2 T^2}{(4\pi r)^2} 2\pi A \frac{\frac{1}{2}\eta_z^2 T^2}{\left[\left(\frac{\sqrt{2}}{2}\eta_z T\right)^4 + (\eta_{xy}T)^2\right]^{3/2}} \tag{6.30}$$

is obtained. Equation (6.30) for Brownian surfaces can also be obtained from both Equations 6.24) and (6.25): letting $H = 1/2$, these two equations coincide, and their series terms are recognized to be the Taylor series expansion of the function $(1 + t^2)^{-3/2}$ (see Equation [6.30]).

6.4.3. Marginally Fractal Surfaces (H→1)

The limiting case $H \to 1$ corresponds to a surface with fractal dimension equal to 2—that is, a regular ("marginally" rough fractal) surface. In this case, the topothesy T has no meaning, and the parameter s becomes dimensionless and assumes the meaning of the rms slope of the surface. Starting directly from Equation (6.11) with $H = 1$ and Equation (A.6.8) with $b = \eta_{xy}, u = \eta_z^2 s^2/2$, the result

$$\left\langle |E_{pq}|^2 \right\rangle = \frac{k^2 |E_p|^2 |F_{pq}|^2}{(4\pi r)^2} 2\pi A \frac{1}{\eta_z^2 s^2} \exp\left(-\frac{\eta_{xy}^2}{2\eta_z^2 s^2}\right) \quad (6.31)$$

is obtained. Equation (6.31) can also be obtained by recasting the series in Equation (6.13) as the Taylor expansion of the exponential function.

It is interesting to note that the quoted result is coincident with that obtained in the case of very rough classical surfaces with Gaussian pdf and Gaussian correlation function.

6.4.4. Quasi-Smooth Surfaces ($kT \ll 1$)

The condition $kT \ll 1$ (quasi-smooth surface) implies that the mean-square surface slope is very small (see Chapter 3), with $\tau_{\min} \propto \lambda$. In this case, the condition in Equation (6.27) is fulfilled, but for directions around the specular one, and use of Equation (6.25) is appropriate. Considering only the first term of the series and using Equations (3.25), it turns out that

$$\left\langle |E_{pq}|^2 \right\rangle = \frac{k^2 |E_p|^2 |F_{pq}|^2}{(4\pi r)^2} A\eta_z^2 \frac{S_0}{\eta_{xy}^\alpha} = \frac{k^2 |E_p|^2 |F_{pq}|^2}{(4\pi r)^2} A\eta_z^2 W(\eta_{xy}). \quad (6.32)$$

This result is analogous to the one obtained for classical surfaces with small roughness.

6.5. Backscattering Coefficient

In remote-sensing applications, a monostatic radar configuration is often employed, so that it is useful to evaluate the backscattering coefficient, usually referred to as normalized radar cross section, NRCS, of the surface:

$$\sigma_{pq}^0 = \frac{4\pi r^2 \left\langle |E_{pq}|^2 \right\rangle}{A |E_p|^2}, \quad (6.33)$$

with coincident incidence and scattering directions:

$$\begin{cases} \vartheta_i = \vartheta_s = \vartheta \\ \varphi_s = \pi \end{cases}. \quad (6.34)$$

6.5. Backscattering Coefficient

Accordingly,

$$\begin{cases} \eta_x = 2k \sin \vartheta \\ \eta_y = 0 \\ \eta_z = -2k \cos \vartheta \end{cases}, \quad (6.35)$$

and

$$\begin{cases} F_{hh}(\vartheta) = -2R_h(\vartheta) \cos \vartheta \\ F_{hv}(\vartheta) = F_{vh}(\vartheta) = 0 \\ F_{vv}(\vartheta) = 2R_v(\vartheta) \cos \vartheta \end{cases}, \quad (6.36)$$

where $R_p(\vartheta)$, with $p = h, v$, is the Fresnel reflection coefficient of the mean plane.

Use of Equations (6.33) through (6.36) in Equations (6.24) and (6.25) leads to

$$\sigma_{pp}^0 = \frac{|R_p(\vartheta)|^2 k^2 T^2 \cos^2 \vartheta}{H} \sum_{n=0}^{\infty} \frac{(-1)^n \Gamma\left(\frac{n+1}{H}\right)}{(n!)^2} \frac{(kT \sin \vartheta)^{2n}}{(\sqrt{2}kT \cos \vartheta)^{\frac{2n+2}{H}}}, \quad (6.37)$$

which is convergent for $0 < H \leq 1/2$, and

$$\sigma_{pp}^0 = |R_p(\vartheta)|^2 k^2 T^2 \cos^2 \vartheta H \sum_{n=1}^{\infty} \frac{(-1)^{n+1} n \Gamma(1+nH)}{n! \Gamma(1-nH)}$$

$$\cdot \frac{(\sqrt{2}kT \cos \vartheta)^{2n}}{(kT \sin \vartheta)^{2nH+2}}, \quad (6.38)$$

which is convergent for $1/2 \leq H < 1$.

Backscattering in the specular direction occurs only in the case of normal incidence, $\vartheta = 0$; in this case, results are polarization independent; Equation (6.37) provides

$$\sigma^0 = \frac{|R(0)|^2 k^2 T^2}{H} \frac{\Gamma\left(\frac{1}{H}\right)}{(\sqrt{2}kT)^{2/H}}, \quad (6.39)$$

which, in the case of a perfectly conducting surface, reduces to the expression that can be found for classical surfaces.

In all the remaining special cases considered in Section 6.4, both Equations (6.37) and (6.38) provide the same results for the backscattering coefficient, as detailed in the following.

For the Brownian surface (Equation [3.7]):

$$\sigma_{pp}^0 = \frac{|R_p(\vartheta)|^2 (\sqrt{2kT}\cos\vartheta)^4}{\left[(\sqrt{2kT}\cos\vartheta)^4 + (2kT\sin\vartheta)^2\right]^{3/2}}. \tag{6.40}$$

For the marginally fractal surface:

$$\sigma_{pp}^0 = \frac{|R_p(\vartheta)|^2}{2s^2} \exp\left(-\frac{\tan^2\vartheta}{2s^2}\right). \tag{6.41}$$

Finally, for the quasi-smooth surface:

$$\sigma_{pp}^0 = \frac{4|R_p(\vartheta)|^2 k^4 \cos^4\vartheta}{\pi} \frac{S_0}{(2k\sin\vartheta)^\alpha}$$

$$= \frac{4|R_p(\vartheta)|^2 k^4 \cos^4\vartheta}{\pi} W(2k\sin\vartheta). \tag{6.42}$$

This last result is analogous to the one obtained for classical surfaces with small roughness.

6.6. Validity Limits

Application of the above-mentioned formulation to the problem of scattering from natural surfaces is subject on one side to the adequacy of the employed surface model, and on the other side to the validity of the Kirchhoff and small-slope approximations. Adequacy of the surface model is examined first.

A preliminary consideration is in order. The formal expression of the expected value of the scattered power density is given by Equation (6.11). It has been noted that the structure of the integrand is such that the main contribution to the integral is essentially provided by a finite range of t values. Accordingly, it is important to assess if the *range of fractalness* includes above mentioned range. To examine this problem, it is convenient to preliminarily analyze the function

$$\tau \exp\left[-\frac{1}{2}k^2 T^2 (\cos\vartheta_i + \cos\vartheta_s)^2 \left(\frac{\tau}{T}\right)^{2H}\right], \tag{6.43}$$

6.6. Validity Limits

which appears in the argument of the integral of Equation (6.11). It can be easily demonstrated that Equation (6.43) exhibits a maximum for

$$\bar{\tau} = T \left(\frac{1}{\sqrt{H}kT(\cos\vartheta_i + \cos\vartheta_s)} \right)^{\frac{1}{H}}. \tag{6.44}$$

This value $\bar{\tau}$ is proportional to $\lambda^{1/H}$. Scale lengths much smaller or much larger than $\bar{\tau}$ do not appreciably contribute to the integrand function in Equation (6.11), and consequently to the generation of the scattered field. In particular, if $H > 0.5$, the values of the function in Equation (6.43) are much smaller than its maximum whenever $\tau < \bar{\tau}/10$ or $\tau > 4\bar{\tau}$; hence, only the interval $[\frac{\bar{\tau}}{10} \div 4\bar{\tau}]$ appreciably contributes to the scattering process. At microwave frequencies, most natural surfaces exhibit a range of fractalness that includes the above-mentioned interval: then natural surfaces can be modeled as fractals at the scales that significantly contribute to the scattering process.

The validity limits of Kirchhoff and small-slope approximations, expressed in terms of fractal parameters, can now be assessed. As already stated, Kirchhoff approximation holds if the surface mean radius of curvature is much greater than the wavelength, and the small-slope approximation is valid if rms slope is much smaller than unity. However, radius of curvature and rms slope are not well defined for a "mathematical" fBm surface. For a "physical" fBm surface, these quantities are related to τ_m, t_M, T, H, as indicated in Table 3.3.

Using the results reported in Table 3.3, condition on rms slope much smaller than unity, $\sigma'^2 \ll 1$, becomes

$$\sigma'^2 = 2^{2H-1}\Gamma^2(1+H)\frac{\sin(\pi H)}{\pi(1-H)}\left[\left(\frac{T}{\tau_m}\right)^{2-2H} - \left(\frac{T}{\tau_M}\right)^{2-2H}\right] \ll 1 \tag{6.45}$$

which, by setting $\tau_m = \bar{\tau}/10$, neglecting the marginal contribution to σ'^2 provided by τ_M, and using Equation (6.44), can be rewritten as

$$10 \cdot 5^{1-2H}\Gamma^2(1+H)\frac{\sin(\pi H)}{\pi(1-H)}\left[\sqrt{H}kT(\cos\vartheta_i + \cos\vartheta_s)\right]^{\frac{2-2H}{H}} \ll 1. \tag{6.46}$$

Furthermore, the condition on the radius of curvature much greater than the wavelength, $\frac{1}{K} \gg \lambda$, gives

$$\frac{k^2}{K^2} \cong \frac{k^2}{\sigma''^2} \cong 2^{1-2H} \frac{\pi(2-H)}{\Gamma^2(1+H)\sin(\pi H)} \frac{k^2 T^2}{\left[\left(\frac{T}{\tau_m}\right)^{4-2H} - \left(\frac{T}{\tau_M}\right)^{4-2H}\right]} \gg 1 \quad (6.47)$$

which, as before by setting $\tau_m = \bar{\tau}/10$, neglecting the marginal contribution to σ''^2 provided by τ_M, and using Equation (6.44), can be rewritten as

$$10^3 \cdot 5^{1-2H} \Gamma^2(1+H) \frac{\sin(\pi H)}{\pi(1-H)} \frac{1}{(kT)^2} \left[\sqrt{H}kT(\cos\vartheta_i + \cos\vartheta_s)\right]^{\frac{4-2H}{H}} \ll 1. \quad (6.48)$$

Equations (6.46) and (6.48) provide approximate guidelines to test the validity of the scattering formulation. More-precise validity limits could be determined—for instance, by comparing theoretical results with numerical simulation of scattering from fractal surfaces.

6.7. Influence of Fractal and Electromagnetic Parameters over the Scattered Field

In this section, the influence of fractal parameters, T and H, on the scattered field is analyzed; for the sake of completeness, the influence of the electromagnetic wavelength is also discussed. The scattering surface is illuminated from the direction $\vartheta_i = \frac{\pi}{4}$, whereas the scattered field is displayed for any ϑ_s in the plane identified by $\varphi_s = 0$; this suggests to assume $-\frac{\pi}{2} \leq \vartheta_s \leq \frac{\pi}{2}$.

The effect of the fractal parameters on the scattered field is first briefly described in a qualitative manner. Then quantitative assessment via Equations (6.24) and (6.25) is graphically presented in Figures 6.3 through 6.6. The presence of the factor $F_{pq}(\cdot)$, which depends on the electromagnetic parameters and field polarizations, is omitted: the reason is that it also appears if classical surfaces are considered, it does not include any dependence on the fractal parameters, and it is not relevant to this discussion. Moreover, the scattered-power density in each figure is normalized to the value of the scattered-power density radiated by the reference surface in the specular direction.

6.7. Fractal and Electromagnetic Parameters over the Scattered Field

The reference scattering surface is characterized by the fractal parameters $S_0 = 0.003076$ m^{2-2H} and $H = 0.8$. These parameters are typical for natural surfaces on the Earth when illuminated by a microwave instrument. Moreover, with the chosen parameters, this reference fBm process holds the same fractal parameters that are held by the WM reference surface (with $\nu = e$) presented in Chapter 3 and used in Chapter 5. Illumination and reference-surface parameters are reported in Table 6.1. The corresponding scattered normalized power density is displayed in Figure 6.3. Then, in any subsequent subsection, only one of the surface fractal parameters is changed within a reasonable and significant range; its specific contribution to the scattered field is graphically displayed. The plot in Figure 6.3 shows that the maximum field scattered by the reference surface is attained in the

TABLE 6.1 Parameters relevant to the illumination conditions and to the reference surface considered in Figure 6.2 and used to compare results in Figures 6.3 through 6.6. These parameters provide a reference surface corresponding to that employed in Chapter 5, where the WM model is used.

Incidence angle (ϑ_i)	45°
Illuminated area (X, Y)	(1 m, 1 m)
Electromagnetic wavelength (λ)	0.1 m
Hurst exponent (H)	0.8
Spectral amplitude (S_0)	0.003076 m^{2-2H}

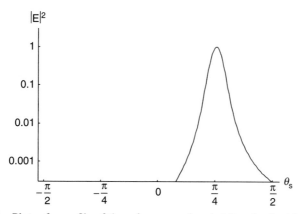

FIGURE 6.3 Plots of a profile of the reference surface holding the fractal parameters $H = 0.8$ and $S_0 = 0.003076$ m^{2-2H} along with the corresponding normalised scattered power density. Illumination and conditions and surface parameters for the reference surface are reported in Table 6.1.

188   6 ◊ Scattering from Fractional Brownian Surfaces: Physical-Optics Solution

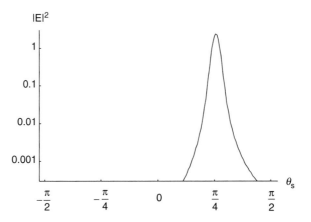

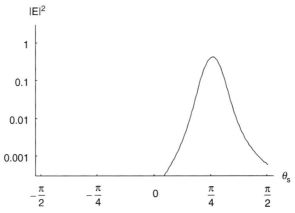

FIGURE 6.4   Modified versions of the reference surface are considered. As in Figure 6.3 corresponding normalized scattered power densities are depicted. top) $S_0 = 0.003076/2$ m^{2-2H}; bottom) $S_0 = 0.003076 \cdot 2$ m^{2-2H}.

specular direction. It is also evident that the average value evaluated for the scattered-power density leads to a very smooth graph.

6.7.1.   The Role of the Spectral Amplitude

The spectral amplitude, S_0, directly influences the surface roughness. In Figure 6.4, two graphs are reported relevant to the power density scattered by surfaces characterized by different values of the spectral amplitude

6.7. Fractal and Electromagnetic Parameters over the Scattered Field

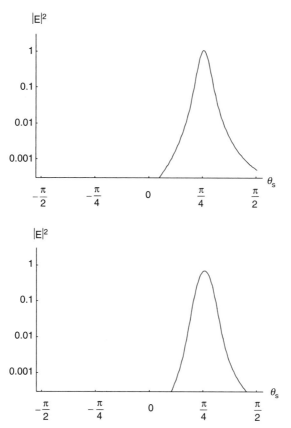

FIGURE 6.5 Modified versions of the reference surface are considered. As in Figure 6.3 corresponding normalized scattered power densities are depicted. top) $H = 0.7$; bottom) $H = 0.9$.

$S_0 = 0.003076/2$ m^{2-2H} and $S_0 = 0.003076 \cdot 2$ m^{2-2H}—one-half and double, respectively, that of the value selected for the reference surface.

Comparison of Figures 6.3 and 6.4 shows that the smaller the spectral amplitude S_0, the narrower the power-density diagram around the specular direction, and the higher the power density scattered in the specular direction.

6.7.2. The Role of the Hurst Exponent

As shown in Chapter 3, the Hurst coefficient is related to the fractal dimension. The lower H, the higher the fractal dimension.

190 6 ◊ Scattering from Fractional Brownian Surfaces: Physical-Optics Solution

In Figure 6.5, two graphs are reported, relevant to surfaces with higher ($H = 0.9$) and lower ($H = 0.7$) values of the Hurst coefficient with respect to the intermediate one ($H = 0.8$) reported in Figure 6.3 and relevant to the reference surface. It is recalled here that changing the value of H while leaving the numerical value of S_0 constant corresponds to considering the amplitude of the power-density spectrum at unitary wave number to be constant.

Comparison of Figures 6.3 and 6.5 shows that the smaller the Hurst coefficient, the narrower the power-density diagram around the specular direction, and the higher the power density scattered in the specular direction.

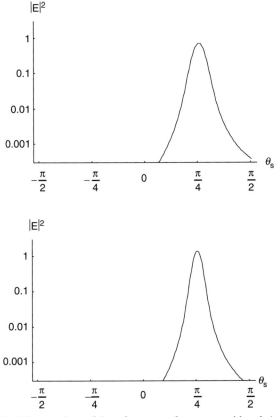

FIGURE 6.6 Modified versions of the reference surface are considered. As in Figure 6.3 corresponding normalized scattered power densities are depicted. top) $\lambda = 0.05$ m; bottom) $\lambda = 0.2$ m.

6.7.3. *The Role of the Electromagnetic Wavelength*

The scattered field evaluated by means of the PO solution and relevant to different values of λ is reported in Figure 6.6.

In Figure 6.6, two graphs are reported, relevant to electromagnetic wavelengths with lower ($\lambda = 0.05$ m) and higher ($\lambda = 0.2$ m) values with respect to the intermediate one ($\lambda = 0.1$ m) reported in Figure 6.3 and relevant to the reference surface. Visual inspection of Equations (6.24) and (6.25) shows that the dependence of the scattered field on the electromagnetic wavelength can be also employed to study the dependence on the topothesy, T.

Comparison of Figures 6.3 and 6.6 shows that the smaller the electromagnetic wavelength, the broader the power-density diagram around the specular direction, and the lower the power density scattered in the specular direction.

6.8. References and Further Readings

As far as classical surfaces are concerned, electromagnetic scattering evaluated by means of the PO solution to the KA is conveniently reported in works by Tsang, Kong, and Shin (1985) and by Ulaby, Moore, and Fung (1982). Falconer (1990) devotes a chapter to fBm surfaces along with relevant definitions and properties. Fundamentals on the convergence of series similar to those discussed in this chapter are presented in a study by Prudnikov, Brychkov, and Marichev (1990). The list of papers written by the authors of this book and concerning the subject of this paper is found in Appendix C.

CHAPTER 7

Scattering from Weierstrass-Mandelbrot Profiles: Extended-Boundary-Condition Method

7.1. Introduction and Chapter Outline

In this chapter, the extended-boundary-condition method (EBCM) is employed with the Weierstrass-Mandelbrot (WM) fractal function to model the surface and solve the electromagnetic-scattering problem. Guidelines of the EBCM have been given in Chapter 4. A key point of the procedure is the property of the WM to be an almost-periodic function. This allows us to generalize mathematical techniques employed for periodic structures, and to express the field by means of a superposition of Floquet modes. The procedure is devised for the general case of dielectric surfaces. In order to simplify the analytic derivation and the employed formalism, the surface profile is considered with a one-dimensional topological dimension. Criteria for assessing the validity of the method are discussed and provided.

As far as classical surfaces are concerned, EBCM is applied to evaluate the scattered field from periodic-surface profiles. For fractal models, the analytic evaluation of the scattered field relies on the property of the WM to be an almost-periodic function, which leads to generalizing the techniques employed in the study of the scattering from periodic interfaces. In particular, it is possible to generalize the Floquet theory, which allows expressing scattered and transmitted electromagnetic fields as a superposition of modes, whose directions of propagation are evaluated by means of the grating equation, and whose amplitudes are computed by means of the EBCM.

The method effectiveness is tested by means of two criteria. The first one is based on energy-balancing considerations, and provides the truncation criterion for the resulting matrices; the second one is based on graphical considerations on scattered-field radiation diagrams.

EBCM solution for the scattered field from a WM profile does not require any surface-roughness constraint that would restrict the solution limits of validity: the obtained solution applies in principle to any surface roughness. However, the surface scattered field is written in terms of an infinite series of amplitude coefficients, whose computation would imply solving a linear system of infinite rank. Numerical solution for the scattering amplitude coefficients calls for an approximate problem of finite rank: to avoid significant degradation of the scattered-field evaluation, this approximate solution needs a profound discussion on the scattered modes and their relative contribution to the whole exact scattered fields. The discussion must be focused on surface-profile parameters, so as to obtain accurate numerical solutions for the scattered field.

This chapter is organized as follows. In Section 7.2 the WM one-dimensional geometric-profile model is presented: the analytic expression for the profile is supported by the required WM parameters settings. Setup of EBCM, in terms of integral equations for a polarized field incident on a prescribed surface profile, is presented in Section 7.3. Surface-fields evaluation is reported in Section 7.4. Incident-, scattered-, and transmitted-field expansions are introduced in Section 7.5. The problem is set in matrix form in Section 7.6 and the analytic derivation of EBCM solution for the scattered field is carried out in Sections 7.7. Section 7.8 deals with the rank order of the matrices involved in the evaluation of the scattering amplitude coefficients. Scattered field modes superposition is discussed in Section 7.9, which also includes relevant consideration of mode truncation and accuracy criteria for numerical solution. Some meaningful numerical examples are reported in Section 7.10. Key references and suggestions for further readings are listed in Section 7.11.

7.2. Profile Model

In this section, the employed geometric-fractal model for a topologically one-dimensional natural-surface profile is presented. Reference is made to Chapter 3, where the two-dimensional topological WM surface model is presented.

Consider the band-limited WM function $z(x)$, which is appropriate to describe natural profiles:

$$z(x) = B \sum_{p=0}^{P-1} C_p \nu^{-Hp} \sin(\kappa_0 \nu^p x + \Phi_p), \qquad (7.1)$$

where B is the vertical-height profile-scaling factor; C_p and Φ_p are random variables that account for the amplitude and the phase behavior of each tone; κ_0 is the wave number of the fundamental component; $\nu > 1$, irrational, is the seed of the geometric progression that accounts for the distribution of the spectral components of the surface; P is the number of tones employed to describe the surface; and H is the Hurst coefficient. Hence, the topologically one-dimensional WM spectrum is formed by a set of P discrete spectral components spaced according to a ν^p law and whose amplitudes follow a ν^{-Hp} law. As in the two-dimensional case reported in Chapter 3, the lower the value of ν, the closer the WM function approximates a fractional Brownian motion (fBm) whose spectrum exhibits a continuous shape.

The main difference with respect to the two-dimensional case relies on the relation between the Hurst exponent and the fractal dimension. It can be demonstrated that the function given in Equation (7.1) belongs to the class of band-limited (physical) fractals whose fractal dimension is $D = 2 - H$. Note that the function in Equation (7.1) is also an almost-periodic function, as formalized by Besicovitch. This property is fundamental in order to employ the generalized Floquet theory for studying the field scattered from fractally corrugated profiles.

As in the two-dimensional case (see Section 3.6.2), use of band-limited WM surfaces is physically justified because any scattering measurement is limited to a finite set of scales. Let X represent the antenna footprint over the surface. The lowest spatial frequency of the surface, $\kappa_0/2\pi$, is linked to the footprint, possibly through an appropriate safety factor $\chi_1 \in (0, 1]$, and its upper spatial frequency $\kappa_0 \nu^{P-1}/2\pi$ is related to the electromagnetic wavelength λ, possibly through another appropriate safety factor $\chi_2 \in (0, 1]$.

Accordingly, it is possible to set

$$\kappa_m = \kappa_0 = \frac{2\pi \chi_1}{X}, \qquad (7.2)$$

and

$$\kappa_M = \kappa_0 \nu^{(P-1)} = \frac{2\pi}{\chi_2 \lambda}. \qquad (7.3)$$

Definitions (7.2) and (7.3) are the equivalent of Equations (3.48) and (3.49) in the topological two-dimensional case, and can be combined to provide the number of tones, $P \in \mathbb{N}$, in terms of relevant model and scattering parameters:

$$P = \left\lceil \frac{\ln(X/\chi_1 \chi_2 \lambda)}{\ln \nu} \right\rceil + 1, \qquad (7.4)$$

where $\lceil \cdot \rceil$ stands for the ceiling function defined so as to take the upper integer of its argument.

7.3. Setup of the Extended-Boundary-Condition Method

In this section, setup of EBCM is presented: this is done to accomplish the specific case at hand, consisting of an incident plane wave impinging on a one-dimensional WM dielectric fractal profile described by means of Equation (7.1). This examination is propaedeutic to evaluating, in the next sections, scattered and transmitted electromagnetic fields.

Hence, in the following discussion the one-dimensional integral equations are set up: decomposition of the incident field onto perpendicular and parallel polarizations allows dealing with scalar integral equations in the unknown (scalar) field ψ. For perpendicular (TE) polarization, ψ stands for the electric field that is aligned along the y-axis; conversely, for parallel (TM) polarization, ψ stands for the magnetic field that is aligned along the y-axis. Finally, incident- and surface-field expansions are presented.

7.3.1. Incident Field

For a single-plane-wave incidence (see Figure 7.1),

$$\psi^{(i)}(\mathbf{r}) = A \exp(-i\mathbf{k_i} \cdot \mathbf{r}), \qquad (7.5)$$

7.3. Setup of the Extended-Boundary-Condition Method

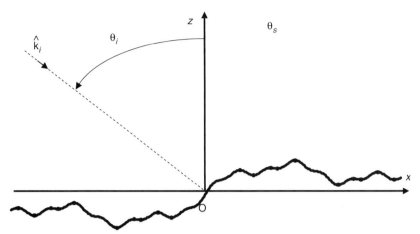

FIGURE 7.1 Geometry of the scattering problem. Cartesian and polar reference systems relevant to the scattering surface are depicted.

where A is an amplitude coefficient, and $\mathbf{k_i} = k_{ix}\hat{\mathbf{x}} + k_{iz}\hat{\mathbf{z}}$ is the incident wave-number vector, with $k_{ix} = k_1 \sin\theta_i$, $k_{iz} = -k_1 \cos\theta_i$, θ_i being the incidence angle.

Extension to more than a single plane wave can be accomplished by means of plane-wave expansion.

7.3.2. Integral Equations

By exploiting the assumption of Sections 7.2 and 7.3.1, the integral equations are written with reference to the cylindrical case that involves a topologically one-dimensional x-dependent-only surface profile, S; separation of TE and TM polarizations allows referring only to the y-component of either the electric or the magnetic field, respectively. This is done in the following paragraphs.

As illustrated in Chapter 4, EBCM is based on the equivalence theorem, and fixes two sets of equivalent sources, which appear inside the integrals in Equations (7.6) through (7.9), and are related to the tangential fields on the scattering-surface profile. Each set of sources radiates in a homogeneous medium holding the same electromagnetic parameters of the upper (subscript 1) or lower (subscript 2) medium. In the case at hand, the first set radiates the electromagnetic scattered field above the surface profile:

$$\psi^{(i)}(\mathbf{r}) + \int_S dS' \left[\psi_1(\mathbf{r}')\hat{\mathbf{n}} \cdot \nabla' g_1(\mathbf{r},\mathbf{r}') - g_1(\mathbf{r},\mathbf{r}')\hat{\mathbf{n}} \cdot \nabla' \psi_1(\mathbf{r}') \right]$$
$$= \psi_1(\mathbf{r}), \quad z > z'(x'), \tag{7.6}$$

and deletes the incident field below the surface profile:

$$\psi^{(i)}(\mathbf{r}) + \int_S dS' \left[\psi_1(\mathbf{r}')\hat{\mathbf{n}} \cdot \nabla' g_1(\mathbf{r},\mathbf{r}') - g_1(\mathbf{r},\mathbf{r}')\hat{\mathbf{n}} \cdot \nabla' \psi_1(\mathbf{r}') \right]$$
$$= 0, \quad z < z'(x'). \tag{7.7}$$

The second set radiates a null field above the surface profile:

$$\int_S dS' \left[\psi_2(\mathbf{r}')\hat{\mathbf{n}} \cdot \nabla' g_2(\mathbf{r},\mathbf{r}') - g_2(\mathbf{r},\mathbf{r}')\hat{\mathbf{n}} \cdot \nabla' \psi_2(\mathbf{r}') \right] = 0, \quad z > z'(x') \tag{7.8}$$

and the electromagnetic transmitted field below the surface profile:

$$\int_S dS' \left[\psi_2(\mathbf{r}')\hat{\mathbf{n}} \cdot \nabla' g_2(\mathbf{r},\mathbf{r}') - g_2(\mathbf{r},\mathbf{r}')\hat{\mathbf{n}} \cdot \nabla' \psi_2(\mathbf{r}') \right]$$
$$= \psi_2(\mathbf{r}), \quad z < z'(x'). \tag{7.9}$$

In Equations (7.6) through (7.9), $\psi^{(i)}(\mathbf{r})$ is the incident field; $\psi_1(\mathbf{r})$ and $\psi_2(\mathbf{r})$ are the total fields above and below the surface, respectively; $\mathbf{r}$ is the generic point in the space surrounding the scattering surface; $\mathbf{r}'$ is a point on the surface profile; the nabla operator, ∇', operates on the primed variable, $\mathbf{r}'$: formally, $\nabla' = \frac{\partial}{\partial x'}\hat{x} + \frac{\partial}{\partial z'}\hat{z}$; $g_1(\mathbf{r},\mathbf{r}')$ and $g_2(\mathbf{r},\mathbf{r}')$ are the scalar Green's functions in the homogeneous two-dimensional media 1 and 2, respectively: their spectral domain representation is

$$g_1(\mathbf{r},\mathbf{r}') = -\frac{i}{4\pi} \int_{-\infty}^{+\infty} dk_{1x} \frac{1}{k_{1z}} \exp\left[-ik_{1x}(x-x') - ik_{1z}|z-z'|\right], \tag{7.10a}$$

$$g_2(\mathbf{r},\mathbf{r}') = -\frac{i}{4\pi} \int_{-\infty}^{+\infty} dk_{2x} \frac{1}{k_{2z}} \exp\left[-ik_{2x}(x-x') - ik_{2z}|z-z'|\right], \tag{7.10b}$$

7.3. Setup of the Extended-Boundary-Condition Method

where k_1 and k_2 are the field wavenumbers in media 1 and 2, respectively, and

$$k_{1z} = \sqrt{k_1^2 - k_{1x}^2}, \tag{7.11a}$$

$$k_{2z} = \sqrt{k_2^2 - k_{2x}^2}. \tag{7.11b}$$

Direct substitution of Equations (7.10) in Equations (7.6) through (7.9) leads to equations that are, in general, not amenable to any further analytic development due to the presence of the absolute value, $|z - z'|$, which is a function of x and x'. This point simplifies if the validity of each integral equation is limited to an appropriate half-space where the absolute values can be easily managed: in particular, substitution of Equation (7.10) in the first and third integral equations leads to managable analytic forms, provided that the analysis is limited to the half-space defined by $z > \max\left[z'(x')\right] = z'_M$; analogously, the region $z < \min\left[z'(x')\right] = z'_m$ is considered for the second and fourth integral equations.

Due to the continuity of the tangential fields on the scattering surface, the two $\psi(\cdot)$ functions included in the integrands appearing in Equations (7.6) through (7.9) must be related. Let

$$\eta = \begin{cases} \mu_2/\mu_1 & TE \text{ case} \\ \varepsilon_2/\varepsilon_1 & TM \text{ case.} \end{cases} \tag{7.12}$$

Then, continuity of the tangential electric and magnetic fields requires that

$$\hat{\mathbf{n}} \times \psi_2(\mathbf{r}')\hat{\mathbf{y}} = \hat{\mathbf{n}} \times \psi_1(\mathbf{r}')\hat{\mathbf{y}}, \tag{7.13a}$$

$$\hat{\mathbf{n}} \times \left[\nabla' \times \psi_2(\mathbf{r}')\hat{\mathbf{y}}\right] = \eta\,\hat{\mathbf{n}} \times \left[\nabla' \times \psi_1(\mathbf{r}')\hat{\mathbf{y}}\right], \tag{7.13b}$$

upon use of the Maxwell equations.

Making use of Equation (A.7.1) with $\mathbf{A} = \hat{\mathbf{n}}$ and $\mathbf{B} = \psi\hat{\mathbf{y}}$, considering that $\hat{\mathbf{n}} \perp \hat{\mathbf{y}}$ and $\left(\hat{\mathbf{n}} \cdot \nabla'\right)\psi(\mathbf{r}')\hat{\mathbf{y}} = \left[\hat{\mathbf{n}} \cdot \nabla'\psi(\mathbf{r}')\right]\hat{\mathbf{y}}$, then the field-continuity conditions (Equations [7.13]) can finally be expressed in terms of the functions appearing in the integrands in Equations (7.6) through (7.9), thus providing

$$\psi_2(\mathbf{r}') = \psi_1(\mathbf{r}') \tag{7.14a}$$

$$\hat{\mathbf{n}} \cdot \nabla'\psi_2(\mathbf{r}') = \eta\,\hat{\mathbf{n}} \cdot \nabla'\psi_1(\mathbf{r}'). \tag{7.14b}$$

Equations (7.7) and (7.8), together with the continuity conditions (Equation [7.14]), are of the integral type, and can be used to evaluate the surface (scalar) fields $\psi_{(1,2)}(\mathbf{r}')$, $\hat{\mathbf{n}} \cdot \nabla' \psi_{(1,2)}(\mathbf{r}')$ in terms of the incident (known) field $\psi^{(i)}(\mathbf{r})$. Once the surface fields are computed, the total field in all the space can be evaluated via Equations (7.6) and (7.9). This is easily recognized to be the solution to the considered problem: it contains the source field $\psi^{(i)}(\mathbf{r})$, satisfies by construction Maxwell equations, and fulfills continuity conditions. Note, however, that the obtained solution turns out to be intrinsically approximated if Equations (7.7) and (7.8) are enforced only outside the strip $z'_m < z < z'_M$ by the imposed restrictions on the Green's functions.

Equations (7.6) through (7.9) can be rearranged in a more compact form. Let us introduce the following Neumann- and Dirichlet-type surface scalar integrals:

$$I_{D(1,2)}^{\pm} \triangleq \int_{S'} g_{(1,2)}(\mathbf{r}, \mathbf{r}') \hat{\mathbf{n}} \cdot \nabla' \psi_{(1,2)}(\mathbf{r}') dS', \tag{7.15}$$

$$I_{N(1,2)}^{\pm} \triangleq \int_{S'} \psi_{(1,2)}(\mathbf{r}') \hat{\mathbf{n}} \cdot \nabla' g_{(1,2)}(\mathbf{r}, \mathbf{r}') dS'. \tag{7.16}$$

Notation is as follows.

The $+$ and $-$ signs apply for the half-spaces $z > \max [z'(x')] = z'_M$ and $z < \min [z'(x')] = z'_m$, respectively. Subscript D or N refers to the two integrands. In the limit of a perfectly conducting lower space, $I_N^{\pm} = 0$ for TE, and $I_D^{\pm} = 0$ for TM polarization, respectively. In these cases, Equations (7.15) and (7.16) are consistent with Dirichlet and Neumann boundary conditions. Subscripts (1 and 2) indicate the wave number k_1 or k_2 to be used.

The above-mentioned notations allow us to express in a very compact form the eight surface integrals involved in the equations for TE and TM cases.

By means of continuity conditions, Equations (7.15) and (7.16) can be written in terms of only ψ_1:

$$I_{D(1,2)}^{\pm} = \eta \int_{S'} g_{(1,2)}(\mathbf{r}, \mathbf{r}') \hat{\mathbf{n}} \cdot \nabla' \psi_1(\mathbf{r}') dS'. \tag{7.17}$$

$$I_{N(1,2)}^{\pm} = \int_{S'} \psi_1(\mathbf{r}') \hat{\mathbf{n}} \cdot \nabla' g_{(1,2)}(\mathbf{r}, \mathbf{r}') dS'. \tag{7.18}$$

Hence, it turns out that Equations (7.6) through (7.9) read as

$$\psi^{(s)}(\mathbf{r}) = \psi_1(\mathbf{r}) - \psi^{(i)}(\mathbf{r}) = I_{N1}^{+} - I_{D1}^{+}; \quad z > z'_M, \tag{7.19}$$

$$\psi^{(i)}(\mathbf{r}) = -I_{N1}^{-} + I_{D1}^{-}; \quad z < z'_m, \tag{7.20}$$

7.3. Setup of the Extended-Boundary-Condition Method

$$0 = I_{N2}^+ - I_{D2}^+; \quad z > z'_M, \tag{7.21}$$

$$\psi_2(\mathbf{r}) = I_{N2}^- - I_{D2}^-; \quad z < z'_m. \tag{7.22}$$

Equations (7.20) and (7.21) are the integral equations to be solved to determine the unknown surface fields in terms of the incident one and surface parameters; then, use of Equations (7.19) and (7.22) allows evaluation of the scattered and transmitted fields.

To this end, appropriate expansion for unknown surface fields is useful, and this is reported in the following subsection.

7.3.3. Surface-Field Expansions for WM Profiles

In EBCM, Equations (7.19) through (7.22) are transformed in a set of infinite algebraic equations by proper surface-field expansions. These, in turn, obviously depend on the geometric property of the scattering profile. Appropriate surface-field expansion deserves some comments whenever a WM function is employed. Compact form of the surface-field expansion calls for rather involved notations, whose rationale deserves some additional comments.

First of all, it is useful to recall the convenient field expansion whenever a sinusoidal surface—hence, one single tone, with spatial period $2\pi/\kappa_0$—is under analysis. In this case, the surface fields are periodic with period $2\pi/\kappa_0$, except for a linear-phase term $\exp(-ik_{ix}x')$, so that a 1-D Fourier expansion is appropriate. As far as $\psi_1(\mathbf{r})$ is concerned, and also including the Jacobian $\sqrt{1 + (dz'/dx')^2}$ of the transformation from dS' to dx'—that is, $dS' = dx'\sqrt{1 + (dz'/dx')^2}$—the periodic functions to be expanded are actually $\sqrt{1 + (dz'/dx')^2}\psi_1(\mathbf{r}')$ and $\sqrt{1 + (dz'/dx')^2}\hat{\mathbf{n}} \cdot \nabla'\psi_1(\mathbf{r}')$. Hence, it turns out that

$$dS'\psi_1(\mathbf{r}') = dx' \exp(-ik_{ix}x') \sum_{q=-\infty}^{+\infty} \alpha_{N,q} \exp(-iq_0 x'), \tag{7.23}$$

$$dS'\hat{\mathbf{n}} \cdot \nabla'\psi_1(\mathbf{r}') = dx'k \exp(-ik_{ix}x') \sum_{q=-\infty}^{+\infty} \alpha_{D,q} \exp(-iq_0 x'), \tag{7.24}$$

where the $\alpha_{D,q}$ and $\alpha_{N,q}$ are the (unknown) coefficients of the Fourier series expansion for the surface (scalar) fields. On the right-hand side of

Equations (7.23) and (7.24), no explicit dependence on z' appears, because z' is a function of x', via the surface-profile Equation (7.1).

If the surface is obtained by superposition of P sinusoidal functions, each one representing a tone, whose periods are $2\pi/\kappa_0, \ldots, 2\pi/\kappa_{P-1}$, then Equations (7.23) and (7.24) can be generalized. In this case, the surface fields are computed by employing the generalized Fourier series expansion, and can be expressed as follows:

$$dS'\psi_1(\mathbf{r}') = dx' \exp(-ik_{ix}x')$$
$$\sum_{q_0=-\infty}^{+\infty} \sum_{q_1=-\infty}^{+\infty} \cdots \sum_{q_{P-1}=-\infty}^{+\infty} \alpha_{N,q_0,\ldots,q_{P-1}}$$
$$\exp\left(-i(q_0\kappa_0 + q_1\kappa_1 + \cdots + q_{P-1}\kappa_{P-1})x'\right), \quad (7.25)$$

$$dS'\hat{n} \cdot \nabla'\psi_1(\mathbf{r}') = dx' k_1 \exp(-ik_{ix}x')$$
$$\sum_{q_0=-\infty}^{+\infty} \sum_{q_1=-\infty}^{+\infty} \cdots \sum_{q_{P-1}=-\infty}^{+\infty} \alpha_{D,q_0,\ldots,q_{P-1}}$$
$$\exp\left(-i(q_0\kappa_0 + q_1\kappa_1 + \cdots + q_{P-1}\kappa_{P-1})x'\right). \quad (7.26)$$

In Equations (7.25) and (7.26), each term of the expansions can be defined as a surface mode. It is composed of an exponential kernel and a coefficient that depend on P indexes q_j, each one expressing the contribution of the corresponding surface-profile tone. Explicit use of the P indexes $q_0, \ldots, q_{P-1}$ makes the combination of tones contributing to the mode immediately visible. Therefore, it is desirable to keep the P indexes in the formal expression of the field. On the other hand, a more compact notation would be also attractive. In order to achieve both aims, two *string indexes*, **q** and **κ**, can be employed:

$$\tilde{\mathbf{q}} = [q_0, \ldots, q_{P-1}], \quad (7.27)$$
$$\tilde{\boldsymbol{\kappa}} = [\kappa_0, \ldots, \kappa_{P-1}], \quad (7.28)$$

where **q** is a column vector $\in Z^P$, whose elements q_j, in principle, span from $-\infty$ to $+\infty$, and **κ** is the column vector of the tones wavenumbers.

7.3. Setup of the Extended-Boundary-Condition Method

This (formal) choice allows us to write Equations (7.25) and (7.26) in the more compact form:

$$dS'\psi_1(\mathbf{r}') = dx'\exp(-ik_{ix}x')\sum_{\substack{q_j=-\infty \\ j=0,\ldots,P-1}}^{+\infty} \alpha_{N,\mathbf{q}}\exp(-i\tilde{\mathbf{q}}\cdot\boldsymbol{\kappa}x'), \quad (7.29)$$

$$dS'\hat{\mathbf{n}}\cdot\nabla'\psi_1(\mathbf{r}') = dx'k_1\exp(-ik_{ix}x')\sum_{\substack{q_j=-\infty \\ j=0,\ldots,P-1}}^{+\infty} \alpha_{D,\mathbf{q}}\exp(-i\tilde{\mathbf{q}}\cdot\boldsymbol{\kappa}x'), \quad (7.30)$$

thus reducing the formal complexity and allowing an easy reading. In particular, each mode is addressed by its pertinent $\mathbf{q}$ string—that is, by the ordered sequence of the chosen components of the string.

A further comment on the introduced notation is appropriate.

The introduced basic notation is the following one:

$$\sum_{\substack{q_j=-\infty \\ j=0,\ldots,P-1}}^{+\infty} Y_{\mathbf{q}} = \sum_{q_0=-\infty}^{+\infty}\sum_{q_1=-\infty}^{+\infty}\cdots\sum_{q_{P-1}=-\infty}^{+\infty} Y_{q_0,\ldots,q_{P-1}}. \quad (7.31)$$

Accordingly, the vector index is a shorthand notation to indicate multiple summations. Furthermore, each summation is an infinite series, because each element of the vector index— say, q_j—spans (in principle) from $-\infty$ to $+\infty$. However, it is anticipated that the practical implementation of EBCM requires truncation of these series, as discussed in Section 7.9.

If the ratios between all pairs κ_i, κ_j are rational numbers, a κ exists such that $\tilde{\boldsymbol{\kappa}} = [n_0\kappa,\ldots,n_{P-1}\kappa]$, all n_j being integers numbers; in this case, the surface profile is simply periodic with spatial period $2\pi/\kappa$. Then Equations (7.29) and (7.30) can be reduced to Equations (7.23) and (7.24), in agreement with the Floquet theorem.

Conversely, if at least one of the ratios between the pairs κ_i, κ_j is an irrational number, then the surface profile is almost periodic, and the generalized Fourier expansions (Equations [7.29] and [7.30]) must be used. In particular, if the WM function defined by Equation (7.1) is used to describe the natural profile, inspection of the equation suggests that $\tilde{\boldsymbol{\kappa}} = [\kappa_0, \kappa_0\nu, \ldots, \kappa_0\nu^{P-1}]$.

The proposed procedure arises from the description of the surface in terms of a Weierstrass-Mandelbrot function. The almost-periodic nature of the WM (due to its construction as a superposition of tones) allows the generalized Fourier series expansion of the surface fields in terms of modes, each mode made by appropriately combining contributions of all the different surface tones and all their possible interactions. As a consequence, and as shown in the following sections, also the scattered field turns out to be expressed as a superposition of modes: again, each scattered mode is due to the mutual interaction of all the tones of the surface.

7.4. Surface-Fields Evaluation

Analytic evaluation of the scattered field via Equations (7.19) through (7.22) implies solving a set of integral equations that involve Dirichlet ($I^{\pm}_{D(1,2)}$)- and Neumann ($I^{\pm}_{N(1,2)}$)-type integrals defined in Equations (7.15) and (7.16). The rationale for accomplishing this task is depicted in this section; analytic details concerning the evaluation of the Dirichlet- and Neumann-type integrals are found in Appendix 7.A, where the above-mentioned integrals are analytically solved in closed form, thus leading to a similarly closed form for the scattered and transmitted fields. For the sake of simplicity, to avoid considering very complicated formal expressions, the considered surface profile is assumed to be indefinite, the x' variable extending over the whole real axis: this is acceptable for surfaces large in terms of the incident electromagnetic wavelength. The case of x' extending on a finite extension of the real axis is discussed later.

Equations (7.19) through (7.22), which define the surface fields and allow computation of the scattered and transmitted field, can be rewritten upon use of the Dirichlet- and Neumann-type integral evaluations provided in Appendix 7.A:

$$\psi^{(s)}(\mathbf{r}) = \psi_1(\mathbf{r}) - \psi^{(i)}(\mathbf{r}) = I^+_{N1} - I^+_{D1}$$

$$= \sum_{\substack{l_j=-\infty \\ j=0,\ldots,P-1}}^{+\infty} \left(a^+_{N1,\mathbf{l}} - a^+_{D1,\mathbf{l}}\right)\exp(-i\mathbf{k}^+_{\mathbf{l}\mathbf{l}} \cdot \mathbf{r}); \quad z > z'_M$$

(7.32)

$$\psi^{(i)}(\mathbf{r}) = -I^-_{N1} + I^-_{D1}$$

$$= \sum_{\substack{l_j=-\infty \\ j=0,\ldots,P-1}}^{+\infty} \left(-a^-_{N1,\mathbf{l}} + a^-_{D1,\mathbf{l}}\right)\exp(+i\mathbf{k}^-_{\mathbf{l}\mathbf{l}} \cdot \mathbf{r}); \quad z < z'_m,$$

(7.33)

$$0 = I_{N2}^+ - I_{D2}^+ = \sum_{\substack{l_j=-\infty \\ j=0,\ldots,P-1}}^{+\infty} \left(a_{N2,\mathbf{l}}^+ - a_{D2,\mathbf{l}}^+\right)\exp\left(-i\mathbf{k}_{2\mathbf{l}}^+ \cdot \mathbf{r}\right); \quad z > z_M',$$

(7.34)

$$\psi_2(\mathbf{r}) = I_{N2}^- - I_{D2}^- = \sum_{\substack{l_j=-\infty \\ j=0,\ldots,P-1}}^{+\infty} \left(a_{N2,\mathbf{l}}^- - a_{D2,\mathbf{l}}^-\right)\exp\left(+i\mathbf{k}_{2\mathbf{l}}^- \cdot \mathbf{r}\right); \quad z < z_m'.$$

(7.35)

Equations (7.32) through (7.35) are a reformulation of Equations (7.19) through (7.22), with the advantage that all integral operations have been analytically performed. A proper reading of these equations is in order.

The Dirichlet- and Neumann-type integrals have been solved, and the obtained closed-form expressions consist of a modal expansion: in each term of right-hand side of Equations (7.32) through (7.35), a P-infinity number of modes is present, each one specified by the choice of the P-indexes of the $\mathbf{l}$ string formally defined by Equation (7.A.11). Each mode is characterized by the wave vector $\mathbf{k}_{(1,2)\mathbf{l}}^\pm$ (see Equation [7.A.16]). In medium 1 and 2, the direction of propagation of each scattered or transmitted mode, defined by the string $\mathbf{l}$, is given by the (possibly complex) angles provided by Equations (7.A.17) and (7.A.18), respectively: these equations are usually referred to as *grating-modes equations*.

The coefficients a appearing in Equations (7.32) and (7.35) are related, via Equations (7.A.15) and (7.A.29), to the (unknown) coefficients α of the Fourier series expansions for the surface fields (see Equations [7.29] and [7.30]). The further step is to manipulate Equations (7.32) through (7.35) such that the latter coefficients α directly appear into the above-mentioned equations set, and can be evaluated. The pertinent procedure is sketched in the subsequent sections.

7.5. Fields Expansions

To proceed further, examination of Equations (7.32) through (7.35) is in order.

It is noted that the left-hand sides of Equations (7.33) and (7.34), and of Equations (7.32) and (7.35), are not on the same footing. The former are known fields, whereas the latter are unknown. Accordingly, we use the first couple to determine the unknown coefficients; after these coefficients have been computed, we use the second couple to evaluate the scattered and transmitted fields. Before implementing the necessary mathematical manipulations, a preliminary expansion of the (known and unknown) fields is convenient.

The observation that the right-hand-side members of Equations (7.32) through (7.35) consist of superposition of waves characterized by wavevectors $\mp \mathbf{k}_{(1,2)\mathbf{l}}^{\pm}$ suggests that we also expand their left-hand side (also including, for the sake of completeness, the null field) onto the same basis. For the first couple, Equations (7.33) and (7.34), the representation is the following one:

$$\psi^{(i)}(\mathbf{r}) = \sum_{\substack{l_i=-\infty \\ i=0,\ldots,P-1}}^{+\infty} c_\mathbf{l}^- \exp(i\mathbf{k}_{1\mathbf{l}}^- \cdot \mathbf{r}), \quad z < z'_m, \tag{7.36}$$

$$0 = \sum_{\substack{l_i=-\infty \\ i=0,\ldots,P-1}}^{+\infty} c_\mathbf{l}^+ \exp(-i\mathbf{k}_{2\mathbf{l}}^+ \cdot \mathbf{r}), \quad z > z'_M, \tag{7.37}$$

where $c_\mathbf{l}^\pm$ are the expansion coefficients.

For the second couple, Equations (7.32) and (7.35), one similarly gets

$$\psi^{(s)}(\mathbf{r}) = \sum_{\substack{l_j=-\infty \\ j=0,\ldots,P-1}}^{+\infty} b_\mathbf{l}^+ \exp(-i\mathbf{k}_{1\mathbf{l}}^+ \cdot \mathbf{r}), \quad z > z'_M \tag{7.38}$$

$$\psi_2(\mathbf{r}) = \sum_{\substack{l_j=-\infty \\ j=0,\ldots,P-1}}^{+\infty} b_\mathbf{l}^- \exp(-i\mathbf{k}_{2\mathbf{l}}^- \cdot \mathbf{r}), \quad z < z'_m, \tag{7.39}$$

where $b_\mathbf{l}^\pm$ are the expansion coefficients.

The coefficients $c_\mathbf{l}^\pm$ can be immediately evaluated: in particular, inspection of Equations (7.5) and (7.A.12) suggests that, in the case of the single incident plane wave, the field-amplitude coefficients $c_\mathbf{l}^-$ are all zero, but for the string $\mathbf{l} = [0, 0 \ldots 0]$, which provides an amplitude coefficient equal to the constant A. Obviously, inspection of Equation (7.37) provides a zero value for all the null-field coefficients $c_\mathbf{l}^+$.

7.6. EBCM Equations in Matrix Form

As far as the second set of equations, they provide the formal solution for the scattered field in the upper medium 1 (Equation [7.38]), and the transmitted field for the lower medium 2 (Equation [7.39]), in terms of the (unknown) expansion coefficients $b_1^{\pm}$.

The formal procedure to be implemented is now clear. The first set (Equations [7.36] and [7.37]) is substituted in Equations (7.33) and (7.34). A linear system of equations is obtained (see Section 7.6), which is inverted to finally obtain the expansion coefficients, $\alpha_{D,\mathbf{q}}$ and $\alpha_{N,\mathbf{q}}$, of the surface-tangential fields. Then these coefficients are used in Section 7.6 to compute the expansion coefficients, $b_1^{\pm}$, of the scattered and transmitted fields.

The scattered, incident, null, and transmitted fields are expressed as superpositions of a countable infinity of modes, whose amplitudes are provided by coefficients b^+, c^-, c^+, and b^-, respectively. Modes, coefficients and wave numbers are all dependent on the P indexes l_j. Each combination of the P indexes identifies a direction of propagation and the corresponding amplitude of the scattered, incident, and transmitted fields. This is even more evident if the explicit forms instead of the compact ones are used to express the electromagnetic fields. For instance, as far as the scattered field is concerned, Equation (7.38) reads as

$$\psi^{(s)}(\mathbf{r}) = \sum_{l_0=-\infty}^{+\infty} \sum_{l_1=-\infty}^{+\infty} \cdots \sum_{l_{P-1}=-\infty}^{+\infty} b^+_{l_0,l_1,\ldots,l_{P-1}} \exp\left(-i\mathbf{k}^+_{1,l_0,l_1,\ldots,l_{P-1}} \cdot \mathbf{r}\right). \tag{7.40}$$

7.6. EBCM Equations in Matrix Form

Inspection of Equations (7.36) through (7.39) allows us to write Equations (7.32) through (7.35) in a matrix form. This convenient arrangement is reached along the following procedure. Formal results are presented in Section 7.7; their practical use is detailed in Section 7.8.

The first step is to substitute Equations (7.36) through (7.39) into Equations (7.32) through (7.35), so that the coefficients $b_1^{\pm}, c_1^{\pm}$ appearing at the left-hand sides are related to the coefficients $a_1^{\pm}$ appearing at their right-hand sides. The second step is to express the coefficients $a_1^{\pm}$ via the surface-modes expansion coefficients, $\alpha_{D,\mathbf{q}}$ and $\alpha_{N,\mathbf{q}}$, given in Equations (7.A.15) and (7.A.29). To proceed further, it is convenient to slightly modify the surface mode expansion coefficients as follows:

$$\alpha'_{D,\mathbf{q}} = -\frac{i}{4\pi}\alpha_{D,\mathbf{q}}\exp(i\tilde{\mathbf{q}}\cdot\boldsymbol{\Phi}), \tag{7.41}$$

$$\alpha'_{N,\mathbf{q}} = \frac{1}{4\pi}\alpha_{N,\mathbf{q}}\exp(i\tilde{\mathbf{q}}\cdot\boldsymbol{\Phi}), \tag{7.42}$$

which also includes the dependence on the profile tones phase $\boldsymbol{\Phi}$ (see Equation [7.A.7]).

At this stage, by considering Expansions (7.36) through (7.39), each of the Equations (7.32) through (7.35) appears as an identity between two infinite series, each one containing appropriate coefficients times identical exponential terms, which are the only **r**-depending functions. The identities should be verified for any **r** in the allowed zones: accordingly, the corresponding coefficients must be equated to each other.

Then Equations (7.33) and (7.34) attain the following formal matrix form:

$$\begin{cases} \mathbf{Q}_{N1}^{-}\cdot\boldsymbol{\alpha}'_N + \mathbf{Q}_{D1}^{-}\cdot\boldsymbol{\alpha}'_D = \mathbf{c}^{-} \\ \mathbf{Q}_{N2}^{+}\cdot\boldsymbol{\alpha}'_N - \eta\mathbf{Q}_{D2}^{+}\cdot\boldsymbol{\alpha}'_D = \mathbf{0} \end{cases}, \tag{7.43}$$

where right-hand sides and left-hand sides have been exchanged, to comply with the usual formal presentation of a linear system where (matrix) coefficients and (vector) unknowns are on the left-hand side, whereas (vector) known terms appear at the right-hand side.

Equations (7.32) and (7.35) are similarly recast as

$$\mathbf{b}^{+} = \mathbf{Q}_{N1}^{+}\cdot\boldsymbol{\alpha}'_N - \mathbf{Q}_{D1}^{+}\cdot\boldsymbol{\alpha}'_D, \tag{7.44}$$

$$\mathbf{b}^{-} = \mathbf{Q}_{N2}^{-}\cdot\boldsymbol{\alpha}'_N - \eta\mathbf{Q}_{D2}^{-}\cdot\boldsymbol{\alpha}'_D. \tag{7.45}$$

In Equations (7.43) through (7.45), the matrices' known entries are given by

$$\mathbf{Q}_{D(1,2),\mathbf{ql}}^{\pm} = (\pm 1)^{m(\mathbf{l})}(\pm 1)^{m(\mathbf{q})}\frac{k_{(1,2)}}{k_{(1,2)z\mathbf{l}}}\exp\left(-i\tilde{\mathbf{l}}\cdot\boldsymbol{\Phi}\right)$$

$$\prod_{p=0}^{P-1}J_{l_p-q_p}\left(k_{(1,2)z\mathbf{l}}BC_p\nu^{-Hp}\right), \tag{7.46}$$

$$Q^{\pm}_{N(1,2),\mathbf{ql}} = (\pm 1)^{m(\mathbf{l})}(\pm 1)^{m(\mathbf{q})} \frac{k^2_{(1,2)} - k_{xl}k_{x\mathbf{q}}}{k^2_{(1,2)z\mathbf{l}}} \exp\left(-i\tilde{\mathbf{l}} \cdot \boldsymbol{\Phi}\right)$$

$$\prod_{p=0}^{P-1} J_{l_p-q_p}\left(k_{(1,2)z\mathbf{l}}BC_p v^{-Hp}\right). \tag{7.47}$$

As already stated, Equations (7.46) and (7.47) provide the entries of the matrices $\boldsymbol{Q}^{\pm}_D$ and $\boldsymbol{Q}^{\pm}_N$. Each entry is identified by a couple of string indexes, $\mathbf{q}$ and $\mathbf{l}$, as defined in Equations (7.27) and (7.A.11). Each element of these matrices relates a surface-field mode, identified by the string $\mathbf{q}$ (see Equations [7.29] and [7.30]), to a radiated-field mode, identified by the string $\mathbf{l}$ (see Equations [7.36] through [7.39]).

A first observation is that the original integral equations (Equations [7.19] through [7.22]) are now transformed into matrix equations that in principle allow their solution. At this stage, the matrices dimensions are infinite, but this point is dealt with in Section 7.8. A further advantage of above the formulation is that the matrix equations do not involve the variable $\mathbf{r}$ at all. Hence, solution of the equations is greatly simplified. In Equations (7.43), the right-hand sides are known, and the unknowns $\boldsymbol{\alpha}'_D$ and $\boldsymbol{\alpha}'_N$ can be evaluated. Substitution of the obtained values in Equations (7.44) and (7.45) allows computing the amplitude coefficients, $\mathbf{b}^+$ and $\mathbf{b}^-$, of the scattered and transmitted fields.

7.7. Matrix-Equations Solution

Solution of the linear system (Equations [7.43]) is in order.

Formal inversion of the (linear) system (Equation [7.43]) leads to the unknowns $\boldsymbol{\alpha}'_D$ and $\boldsymbol{\alpha}'_N$ evaluation:

$$\boldsymbol{\alpha}'_D = \boldsymbol{W}_1^{-1}(\eta) \cdot \mathbf{c}^- \tag{7.48}$$

$$\boldsymbol{\alpha}'_N = \eta \boldsymbol{Q}_{N2}^{+~-1} \cdot \boldsymbol{Q}_{D2}^{+} \cdot \boldsymbol{W}_1^{-1}(\eta) \cdot \mathbf{c}^- \tag{7.49}$$

where

$$\boldsymbol{W}_1(\eta) = \boldsymbol{Q}_{N1}^{-} \cdot \eta \boldsymbol{Q}_{N2}^{+~-1} \cdot \boldsymbol{Q}_{D2}^{+} + \boldsymbol{Q}_{D1}^{+}. \tag{7.50}$$

Equations (7.48) through (7.50) allow us to express the amplitude coefficients of the surface-fields modes, $\boldsymbol{\alpha}'$, in terms of the incident field, $\mathbf{c}^-$, and the surface-profile parameters (included in the matrices).

Substitution of Equations (7.48) and (7.49) in Equation (7.44) leads to

$$\mathbf{b}^+ = \left(\boldsymbol{Q}_{N1}^+ \cdot \eta \boldsymbol{Q}_{N2}^{+}{}^{-1} \cdot \boldsymbol{Q}_{D2}^+ - \boldsymbol{Q}_{D1}^+ \right) \cdot \boldsymbol{\alpha}'_D = W_2(\eta) \cdot W_1^{-1}(\eta) \cdot \mathbf{c}^-, \quad (7.51)$$

where

$$W_2(\eta) = \boldsymbol{Q}_{N1}^+ \cdot \eta \boldsymbol{Q}_{N2}^{+}{}^{-1} \cdot \boldsymbol{Q}_{D2}^+ - \boldsymbol{Q}_{D1}^+. \quad (7.52)$$

Equations (7.51) and (7.52) allow us to express the amplitudes, $\mathbf{b}^+$, of scattered-field modes in terms of the amplitudes coefficients, $\mathbf{c}^-$, of the incoming field and the surface-profile parameters. Hence, the scattered field is expressed as a countable superposition of modes, whose amplitude coefficients are given by Equation (7.51), and whose directions of propagation are given by Equation (7.A.17).

Substitution of Equations (7.48) and (7.49) in Equation (7.45) leads to

$$\mathbf{b}^- = \left(\boldsymbol{Q}_{N2}^- \cdot \eta \boldsymbol{Q}_{N2}^{+}{}^{-1} \cdot \boldsymbol{Q}_{D2}^+ - \eta \boldsymbol{Q}_{D2}^- \right) \cdot \boldsymbol{\alpha}'_D = W_3(\eta) \cdot W_1^{-1}(\eta) \cdot \mathbf{c}^-, \quad (7.53)$$

where

$$W_3(\eta) = \boldsymbol{Q}_{N2}^- \cdot \eta \boldsymbol{Q}_{N2}^{+}{}^{-1} \cdot \boldsymbol{Q}_{D2}^+ - \eta \boldsymbol{Q}_{D2}^-. \quad (7.54)$$

Equations (7.53) and (7.54) allow us to express the amplitudes, $\mathbf{b}^-$, of transmitted-field modes in terms of the amplitudes, $\mathbf{c}^-$, of the incoming field and the surface-profile parameters. Hence, the transmitted field is expressed as a superposition of modes whose amplitude coefficients are evaluated via Equation (7.54), and whose directions of propagation are given by Equation (7.A.18).

7.8. Matrices Organizations

In Equations (7.51) and (7.53), the amplitude coefficients can be evaluated once the matrices W_1, W_2, W_3 are computed. To this aim, two main problems must be solved: matrices elements ordering and matrices dimensions.

7.8. Matrices Organizations

As far as the ordering problem is concerned, note that, as stated in Section 7.6, each matrix element in Equations (7.43) through (7.54) is defined by the chosen entries of the strings **q** and **l**. String **q** states which mode is considered for the surface-tangential fields (see Equations [7.25] and [7.26]): any choice of the P elements of **q** values corresponds to a combination of tones relevant to the surface profile and identifies a surface-field mode. String **l** states which mode is considered for the scattered, incident, null, or transmitted field (see Equations [7.36] through [7.39]): any choice of the P elements of **l** values identifies a scattered-field mode, whose direction of propagation is given by the grating equations (Equations [7.A.17] and [7.A.18]). Hence, each matrix element in Equations (7.43) through (7.54) is identified by $2P$ indexes: P indexes are employed to define the string **q**, which identifies the matrix row, whereas the other P indexes are employed to define the string **l**, which identifies the matrix column. Then the practical construction of the matrices depends on the rule that is employed to order the elements of the strings **q** and **l**. However, the order of the matrix elements does not affect the evaluation of **b**, provided that the same criterion is adopted for each matrix appearing in Equations (7.43) through (7.54), and consequently for each matrix or vector involved in the overall procedure depicted in Section 7.7.

Matrices appearing in Equations (7.43) through (7.54)—as well as vectors $\mathbf{b}^+, \mathbf{b}^-, \mathbf{c}^+$, and $\mathbf{c}^-$—have no finite dimension because they are generated by varying P indexes between $-\infty$ and $+\infty$. These infinite dimensions are due to the infinite number of terms associated to the generalized Fourier series expansion of the surface field in Equations (7.25) and (7.26). Then it is necessary to truncate the matrices if numerical methods may be used to evaluate the expansion coefficients of surface and scattered fields.

It is important to choose a criterion that allows considering only a finite number of relevant elements of the matrices, without impairing the precise evaluation of the scattering coefficients. To this end, it is preliminarily required to categorize the modes and the matrix elements.

Modes can be categorized according to the value of the strings **l**. A radiated mode is defined of K-th (interaction) order if

$$\sum_{j=0}^{P-1} |l_j| = K. \tag{7.55}$$

A similar definition can be set for the surface modes by referring to the string **q**.

It is convenient to set the finite number of modes employed to evaluate the scattered fields by using only modes whose interaction order is smaller than a prescribed value $K_{\max}$ that is chosen according to a criterion presented in the next section. In other words, only modes characterized by strings **l** and **q** such that

$$\begin{cases} \sum_{j=0}^{P-1} |l_j| \leq K_{\max} \\ \sum_{j=0}^{P-1} |q_j| \leq K_{\max} \end{cases} \tag{7.56}$$

are considered. This criterion truncates the matrices Q to a finite order that increases exponentially with K. As a matter of fact, assume that at the interaction order K, n strings **q** and n strings **l** are generated. Change of the order from K to $K+1$ produces additional strings for each already-existing string: accordingly, the strings-number increase, Δn, is proportional to n, with the resulting exponential pattern.

However, the number of significant entries for each Q matrix is recognized to be smaller than the number identified by Conditions (7.56): for real values of $k_{(1,2)zl}$ (not evanescent modes), which account for the propagation in the far field, the Q entries are significant only if the order of each Bessel function is lower than its argument. Then, for each l_j, only those q_j that satisfy the condition

$$|l_j - q_j| < k_{(1,2)zl} B C_p v^{-Hp} \tag{7.57}$$

(see Equations [7.46] and [7.47]) significantly contribute to the formation of the scattered field. If only the far field must be evaluated, it is possible to approximate to zero the entries of the Q matrices that do not satisfy Equation (7.57). It turns out that only the surface-fields modes identified by string values **q** close enough to the string value **l** (see Equation [7.57]) contribute to the formation of the scattered-far-field mode identified by **l**. It is concluded that, except for very rough surfaces, the matrices whose entries are presented in Equations (7.46) and (7.47) are sparse.

The above results lead to an evaluation of the dimensions of matrices as described in Table 7.1, where N_b is the number of modes of order smaller or equal to $K_{\max}$, $N_p(N_p < N_b)$ is the number of propagating modes in the

TABLE 7.1 Dimensions of the matrices involved in the EBCM approach.

Matrix	Dimensions	Matrix	Dimensions
$\mathcal{Q}_{D1}^+$	$N_p \times N_b$	$\mathcal{Q}_{D2}^+$	$N_b \times N_b$
$\mathcal{Q}_{D1}^-$	$N_b \times N_b$	$\mathcal{Q}_{D2}^-$	$N_T \times N_b$
$\mathcal{Q}_{N1}^+$	$N_p \times N_b$	$\mathcal{Q}_{N2}^+$	$N_b \times N_b$
$\mathcal{Q}_{N1}^-$	$N_b \times N_b$	$\mathcal{Q}_{N2}^-$	$N_T \times N_b$

upper medium, and N_T ($N_T < N_b$) is the number of propagating modes in the lower medium.

7.9. Scattering-Modes Superposition, Matrices Truncation, and Ill-Conditioning

In Section 7.8, the matrices dimensions are set according to an appropriate selected maximum-interaction order; then it is important to establish a criterion that fixes this maximum-interaction order $K_{\max}$, consistent with the required accuracy and calculation-time constraint. The criterion is based on the energy-conservation law.

Define the energy parameter, e, as the sum, normalized to the incident power, of reflected and transmitted powers. For the considered case of the single incident plane wave:

$$e = \frac{1}{|A|^2 \cos \theta_i} \left(\sum_{l=1}^{N_p} |b_l^+|^2 \cos \theta_{1l} + \chi \sum_{l=1}^{N_T} |b_l^-|^2 \cos \theta_{2l} \right), \quad (7.58)$$

wherein

$$\chi = \begin{cases} \dfrac{\sqrt{\mu_1/\varepsilon_1}}{\sqrt{\mu_2/\varepsilon_2}}, & \text{TE case} \\[2ex] \dfrac{\sqrt{\mu_2/\varepsilon_2}}{\sqrt{\mu_1/\varepsilon_1}}, & \text{TM case.} \end{cases} \quad (7.59)$$

If $e = 1$, the energy-conservation law is satisfied: this limiting case is approached as the interaction order is increased. The criterion for the $K_{\max}$

choice is formalized by the following relationships:

$$\left|e_{K_{\max}} - 1\right| < \delta_1, \quad (7.60)$$

$$\left|e_{K_{\max}} - e_{K_{\max}-1}\right| < \delta_2, \quad (7.61)$$

wherein δ_1 and δ_2 are small parameters to be set. Equation (7.60) ensures that the parameter e converges to the desired value, whereas Equation (7.61) is a convergence criterion. Implementation of Equations (7.60) and (7.61) leads to fixing the interaction order $K_{\max}$ at which the calculations can be stopped.

The matrices involved in the numerical computation can be ill-conditioned. This occurs whenever at least one of the singular values of the matrices that must be inverted is much smaller than the others. In this case, small changes in the matrix entries—for instance, due to round-off errors—may cause instability in the numerical matrix inversion, thus leading to incorrect results for the scattered fields. The problem is related to the precision of the computer and of the code employed to perform the computations. Numerical solutions based on singular-value decomposition can be employed to relax this constraint.

Ill-conditioning problems may arise whenever very rough surfaces are taken into account. In this case, the value of B increases, thus increasing the arguments of the Bessel functions (see Equations [7.46] and [7.47]), and the number of modes that significantly contribute to the scattered field (see Equation [7.57]). Then matrices entries and dimensions are increased, resulting in a larger probability of having a large ratio of the largest to the smallest singular value, and the matrices become ill-conditioned.

Ill-conditioning has a physical counterpart that can be individuated by analyzing Equation (7.34) or Equation (7.33).

Equation (7.34) is considered first. This equation enforces a null field in the region $z > z'_M$, and not in the whole (wider) region $z > z'(x', y')$, as it is prescribed by Equation (7.8). This different condition has been introduced in order to employ the simple expression in Equation (7.10b) for the Green's function in the spectral domain: this expression involves a plane-wave expansion that is valid only in the half-space $z > z'_M$. Roughly speaking, if the roughness is small, the difference between the two regions is limited, and it is reasonable to expect that the field is approximately equal to zero also in the region $z'(x', y') < z < z'_M$. If the surface roughness increases—for instance, by increasing B—even if the field in the half-space $z > z'_M$ is null, different and significant fields can exist in the region $z'(x', y') < z < z'_M$, so that the numerical evaluation of surface fields may be unstable. As a matter of fact, very different surface fields may radiate different fields in the region $z'(x', y') < z < z'_M$, but may radiate practically the same null field in the half-space $z > z'_M$—so that they are virtually equivalent from the viewpoint of Equation (7.34), and ill-conditioning arises. Analogous considerations hold for matrix Equation (7.33), which is equivalent to enforce Equation (7.7) only in the half-space $z < z'_m$, and not in all the region $z < z'(x', y')$.

Then matrices ill-conditioning has a practical drawback: the numerical procedure cannot be applied to surface profiles of any (somehow defined) roughness; and for the EBCM, a maximum roughness needs to be considered. This maximum-roughness value cannot be set in general: it obviously depends on the considered situation (surface profile and impinging electromagnetic field), as well as (and less obviously) on the computer and implemented numerical code that are employed to solve the scattering problem.

7.10. Influence of Fractal and Electromagnetic Parameters over the Scattered Field

In this section, the influence of fractal parameters—κ_0, ν, P, B, and H—on the scattered field is analyzed: the WM coefficients are assumed to be not random: $C_p = 1$, $\Phi_p = 0$ for any p. The scattering surface is illuminated from the direction $\vartheta_i = \pi/4$, whereas the scattered field is displayed for any ϑ_s; this suggests assuming $-\frac{\pi}{2} < \vartheta_s \leq \frac{\pi}{2}$.

First, the effect of the fractal parameters on the scattered field is briefly described in a qualitative manner. Then quantitative assessment is graphically presented in Figures 7.2 through 7.9.

A reference case is considered first. Then, in any subsequent subsection, only one of the surface fractal parameters is changed within a reasonable and significant range.

The reference scattering surface is characterized by the fractal parameters listed in Table 7.2. These parameters are typical for natural surfaces on the Earth when illuminated by a microwave instrument. The scattering profile is chosen to be of finite length. This implies that in the series at the right-hand side of Equations (7.32) through (7.35), each exponential term is replaced by a sinc($\cdot$) function, whose maximum is aligned to the direction individuated by the wavevectors in Equations (7.A.16) through (7.A.18)

TABLE 7.2 Parameters relevant to the electromagnetic illumination and to the reference surface of Figure 7.2.

Incidence angle (ϑ_i)	45°
Illuminated area (X)	1 m
Electromagnetic wavelength (λ)	0.1 m
Hurst exponent (H)	0.8
Tone wave-number spacing coefficient (ν)	e
Overall amplitude-scaling factor (B)	0.01 m

for the profile of infinite length. This leads to represent also the scattered, transmitted, incident and null fields in Equations (7.36) through (7.39) as superposition of sinc(·) functions, each sinc(·) exhibiting the maximum field in the directions individuated by Equations (7.A.16) through (7.A.18).

Same normalisations introduced in Section!5.5 are in this section applied: the reason is to allow the comparison with results shown in Section 5.5, thus emphasizing the dependence of the scattered power-density on the fractal parameters. Scattering surface and normalized *scattered power-density* are displayed in Figure 7.2. In this figure and in the following ones, the scattered-power density is normalized to the value of the scattered-power density radiated by the reference surface in the specular direction. Graphs in Figure 7.2 have been obtained with $K_{\max} = 2$. In Figure 7.3, the scattered-power density evaluated for $K_{\max} = 1, 2, 3$ is displayed: it can be appreciated that for the reference surface illuminated under the reference condition reported at the beginning of this section, the scattered power-density is correctly represented by choosing $K_{\max} = 2$. This is confirmed by the corresponding values attained by the e parameter: $e_1 = 0.98637$; $e_2 = 0.99877$; $e_3 = 0.99997$. All graphs in the subsequent subsections have been obtained by setting $\delta_1 = \delta_2 = 0.01$ in Equations (7.60) and (7.61) to determine $K_{\max}$.

Plots in Figure 7.2 show that the maximum field scattered by the reference surface is attained in the specular direction. For angles close to the specular direction, the scattered field approximately exhibits the sinc(·) behavior, showing that the specular mode is the dominant one. The scattered field is arranged in a series of lobes whose mean width is proportional to the ratio between the electromagnetic wavelength and the length of the illuminated area. The overall scattered field is obtained as superposition of the significant scattered modes. According to the considerations reported in Chapter 3 and Section 7.8, 4 tones are necessary to describe the surface, and 24 localized

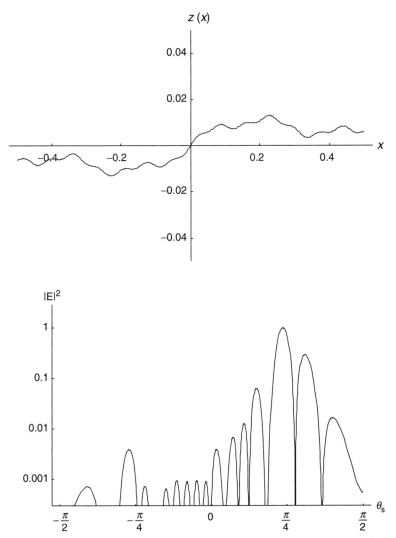

FIGURE 7.2 Plot of the profile, $y = 0$, of the reference surface holding the fractal parameters listed under Table 5.1 along with the graph of the corresponding normalized scattered power density in the plane $y = 0$.

modes have been considered in the numerical evaluation of the scattered-power density.

It is noted that the graph of Figure 7.2 refers to the particular ensemble element of the surface depicted in the same figure. This is convenient for

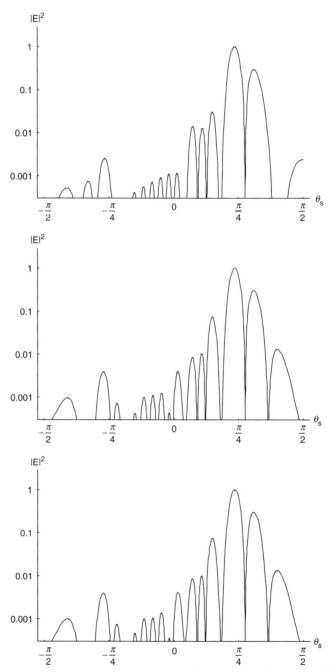

FIGURE 7.3 Plots of the normalized scattered power density by varying the number of interactions: $K_{max} = 1$ top, $K_{max} = 2$ middle, $K_{max} = 3$ bottom figure.

7.10. Influence of Fractal and Electromagnetic Parameters

many applications, but does not explicitly show the statistical behavior of the ensemble itself. To get this additional information, as was already pointed out in Section 1.6.1, the WM model parameters are changed according to their statistics, the associate scattered power-density for each element is computed, and the obtained values are properly processed. For instance, the expected value of the scattered-power density is obtained by averaging thirty-two graphs as that of Figure 7.2, each relative to one element of the surface ensemble. This is shown in Figure 7.4.

7.10.1. The Role of the Fundamental-Tone Wavenumber

As shown in Chapter 3, κ_0 is not an independent fractal parameter. Its value can be set according to the illuminated-surface dimension: as already shown, see Equation (7.2), it is reasonable to set the fundamental-tone wavelength of the order of the maximum length of the surface, with the addition of an appropriate safety factor $\chi_1 \in (0, 1]$, the latter to be set equal to 0.1 in most critical cases. Accordingly:

$$\kappa_0 = \frac{2\pi}{X}\chi_1. \tag{7.90}$$

In the numerical examples in this section, $\chi_1 = 1/\sqrt{2}$ is chosen: this is done to get the same value of κ_0 that was obtained in Chapter 5 for a topologically two-dimensional surface with $X = Y$ under the same illumination conditions.

Instead of κ_0, from the electromagnetic-scattering viewpoint, it is more meaningful to consider the effect induced by a change of the scattering-surface dimensions, X, according to Equation (7.90). The scattered-power density, evaluated in the EBCM and relevant to different values of the surface dimension, is reported in Figures 7.5.

Comparison of results reported in Figures 7.2 and 7.5 shows that the larger the illuminated area—that is, the smaller the value of κ_0—the narrower the radiated lobes, with a corresponding increase in the number of oscillations with ϑ_s. The side-lobes amplitude appears to be unaffected by κ_0. In Figure 7.5a, 3 tones and 15 localized modes have been considered; in Figure 7.5b, 5 tones and 32 localized modes have been considered.

7.10.2. The Role of the Tone Wavenumber Spacing Coefficient

As shown in Section 3.6, ν is the fractal parameter that controls the tone wave-number spacing. Its value sets the degrees of similarity between

7 ◊ Scattering from WM Profiles: EBCM Solution

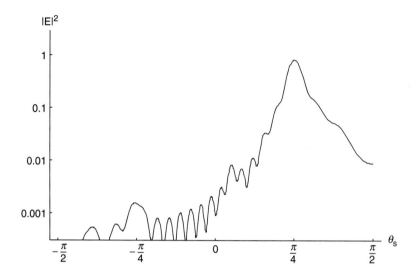

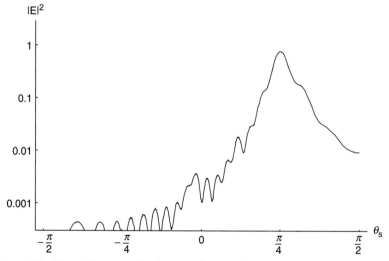

FIGURE 7.4 Plots of the normalized scattered power density relevant to a set of surfaces which have been obtained by randomizing the surface coefficients of the reference surface. Thirty-two realizations have been considered. Top $v = e$; bottom $v = e - 0.5$.

7.10. Influence of Fractal and Electromagnetic Parameters

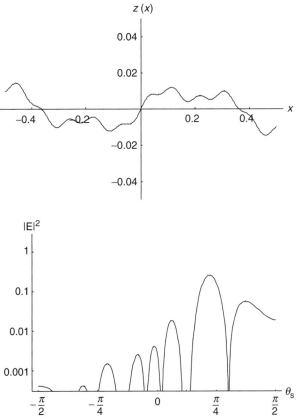

FIGURE 7.5a Modified versions of the reference surface are considered. As in Figure 7.2, surface cuts and corresponding normalized scattered power density graphs are depicted. Case $L_x = L_y = 0.5$ m.

the considered WM function and an appropriate fBm process of appropriate fractal parameters. Its value can be set to any irrational number >1. The scattered power-density relevant to different values of ν is reported in Figures 7.6.

Comparison of Figures 7.2 and 7.6 shows that the tone spacing marginally influences the scattered field. A smoother overall shape for the scattered power-density is obtained for lower values of ν. However, the ν value can greatly influence the number of tones that are required to represent the surface: thus, the number of significant modes, and consequently the computational time of the scattered field, is greatly influenced by the choice of ν.

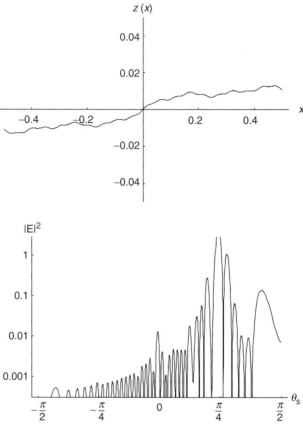

FIGURE 7.5b Modified versions of the reference surface are considered. As in Figure 7.2, surface cuts and corresponding normalized scattered power density graphs are depicted. Case $L_x = L_y = 2$ m.

In Figure 7.6a, 5 tones and 32 localized modes have been considered; in Figure 7.6b, 4 tones and 18 localized modes have been considered. In conclusion, it is confirmed that ν is a fractal parameter that can be set to obtain reliable discrete approximation of the continuous fBm spectral behavior; furthermore, its choice is relevant for the surface-profile representation, but not so much as far as the scattered field is concerned. Accordingly, its value could be set not too close to 1 to allow efficient evaluation of the scattered field.

7.10. Influence of Fractal and Electromagnetic Parameters

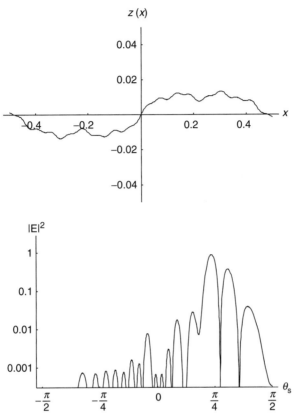

FIGURE 7.6a Modified versions of the reference surface are considered. As in Figure 7.2, surface cuts and corresponding normalized scattered power density graphs are depicted. Case $v = e + 0.5$.

7.10.3. The Role of the Number of Tones

As shown in Chapter 3, P is not an independent fractal parameter. Its value can be set according to the incident wavelength and illuminated-surface dimension: as already shown, see Equation (7.2) through (7.4), it is reasonable to set the surface upper spatial wavenumber of the order of the electromagnetic wavelength, with the addition of an appropriate safety factor $\chi_2 \in (0, 1]$, the latter usually set to 0.1. Accordingly:

$$P = \left\lceil \frac{\ln(X/\chi_1\chi_2\lambda)}{\ln \nu} \right\rceil + 1. \tag{7.91}$$

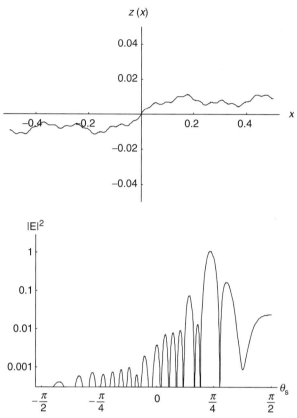

FIGURE 7.6b Modified versions of the reference surface are considered. As in Figure 7.2, surface cuts and corresponding normalized scattered power density graphs are depicted. Case $\nu = e + 0.5$.

In the numerical example in this section, $\chi_2 = 1/\sqrt{2}$ is chosen: this is done to obtain the same value of P that was obtained the examples in Chapter 5 for a topologically two-dimensional surface with $X = Y$ under the same illumination conditions. The number of tones increases logarithmically with the ratio between the surface dimensions to the electromagnetic wavelength. The lower the safety factors or tone spacing, the higher the number of tones.

In view of Equation (7.4) and the results of Section 5.5.3, instead of P, it is more meaningful to consider changes induced by varying the electromagnetic wavelength λ. The scattered power-density evaluated by means of the formula presented in this chapter and relevant to different values of λ is reported in Figures 7.7.

7.10. Influence of Fractal and Electromagnetic Parameters 225

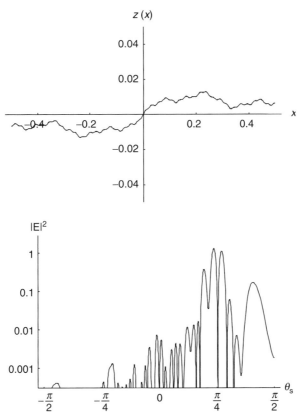

FIGURE 7.7a Modified versions of the reference surface are considered. As in Figure 7.2, surface cuts and corresponding normalized scattered power density graphs are depicted. Case $\lambda = 0.05$ m.

Comparison of Figures 7.2 and 7.7 shows that the larger the number of tones—that is, the smaller the electromagnetic wavelength—the narrower the radiated lobes. For smaller λ, a finer description of the surface is required, which implies a corresponding increase in the number of modes. In Figure 7.7a, 5 tones and 97 modes have been considered; in Figure 7.7b, 3 tones and 15 modes have been considered.

7.10.4. The Role of the Overall Amplitude-Scaling Factor

The overall amplitude-scaling factor B directly influences the surface roughness (Equation [7.1]). Normalized-tone roughness turns out to be

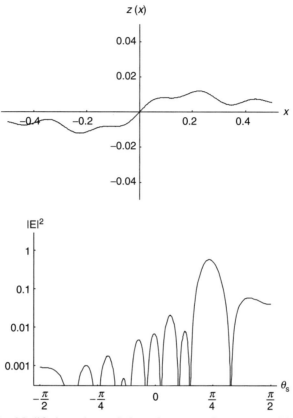

FIGURE 7.7b Modified versions of the reference surface are considered. As in Figure 7.2, surface cuts and corresponding normalized scattered power density graphs are depicted. Case $\lambda = 0.2$ m.

proportional to B. Hence, the number of significant modes, M, increases with B.

In Figures 7.8, two graphs are reported of the scattered-power density relevant to surfaces with lower ($B = 0.005$ m) and higher ($B = 0.02$ m) values of the overall amplitude-scaling factor B with respect to the intermediate one $B = 0.01$ m reported in Figure 7.2 relevant to the reference surface.

Comparison of Figures 7.2 and 7.8 shows that the larger the overall amplitude-scaling factor B, the larger the number of modes that significantly contribute to the scattered field. Accordingly, the scattered power-density

7.10. Influence of Fractal and Electromagnetic Parameters

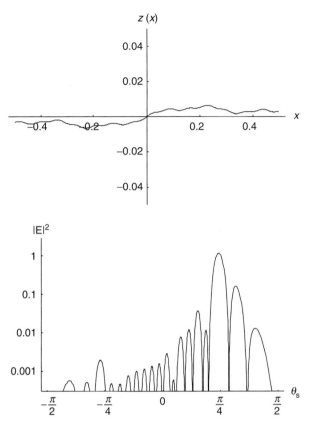

FIGURE 7.8a Modified versions of the reference surface are considered. As in Figure 7.2, surface cuts and corresponding normalized scattered power density graphs are depicted. Case $B = 0.005$ m.

resembles the sinc($\cdot$) behavior for lower B; conversely, for higher B, the sinc($\cdot$) behavior is lost. Note that the number of modes may increase, also varying other fractal parameters; but the disappearance of the sinc($\cdot$) behavior is more evident if B is increased. In Figure 7.8a, 4 tones and 24 modes have been considered; in Figure 7.8b, 4 tones and 122 modes have been considered.

7.10.5. The Role of the Hurst Exponent

As shown in Chapter 3, the Hurst coefficient is related to the fractal dimension. The lower the value of H, the higher the fractal dimension, and the

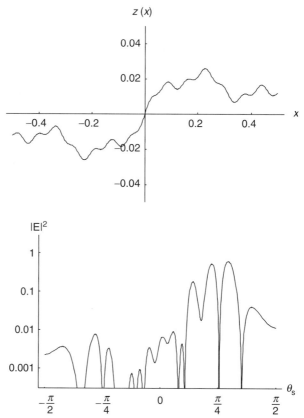

FIGURE 7.8b Modified versions of the reference surface are considered. As in Figure 7.2, surface cuts and corresponding normalized scattered power density graphs are depicted. Case $B = 0.02$ m.

higher the number of significant modes that enter into the evaluation of the scattered field.

In Figures 7.9, two graphs are reported, relevant to surfaces with higher ($H = 0.9$) and lower ($H = 0.7$) values of the Hurst coefficient with respect to the intermediate one ($H = 0.8$) in Figure 7.2 relative to the reference surface.

Comparison of Figures 7.2 and 7.9 shows that the smaller the Hurst coefficient, the higher the lateral lobes. In both Figures 7.9a and 7.9b, 4 tones and 24 modes have been considered.

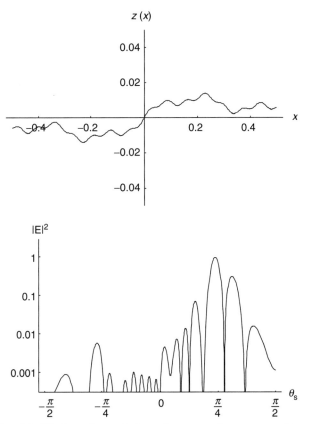

FIGURE 7.9a Modified versions of the reference surface are considered. As in Figure 7.2, surface cuts and corresponding normalized scattered power density graphs are depicted. Case $H = 0.7$ m.

7.11. References and Further Readings

Scattering of waves from periodic surfaces is discussed in works by Chuang and Kong (1981) and by Waterman (1975). Theory supporting EBCM can be found in studies by Kong (1986) and by Tsang, Kong, and Shin (1985). Besicovitch (1932) describes almost-periodic functions. Fractals and useful information to identify their almost-periodic behavior can be found in works by Mandelbrot (1983), Falconer (1990), Voss (1985), and Berry and Lewis (1980). Scattering from WM metallic profiles is presented in a study by Savaidis et al. (1997). Propagation through almost-periodic media is

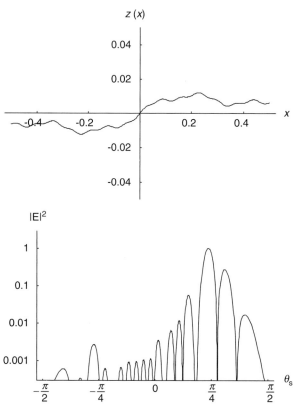

FIGURE 7.9b Modified versions of the reference surface are considered. As in Figure 7.2, surface cuts and corresponding normalized scattered power density graphs are depicted. Case $H = 0.9$ m.

reported by Mickelson and Jaggard (1979). The EBCM studied from the Rayleigh point of view is reported by Tsang, Kong, and Shin and also by Jaggard and Sun (1990a). Floquet modes are presented by Collin (1960). The list of the papers written by the authors of this book and concerning the subject of this chapter is found in Appendix C.

Appendix 7.A Evaluation of the Dirichlet- and Neumann-Type Integrals

Analytic evaluation of the scattered field via Equations (7.19) through (7.22) implies solving a set of integral equations that involve the

Appendix 7.A Evaluation of the Dirichlet- and Neumann-Type Integrals

Dirichlet ($I^{\pm}_{D(1,2)}$)- and Neumann ($I^{\pm}_{N(1,2)}$)-type integrals defined by Equations (7.15) and (7.16). Evaluation of the Dirichlet- and Neumann-type integrals is presented in this section: this leads to a similarly closed-form evaluation of the scattered and transmitted fields.

Solution of the Neumann integral requires the solution of the Dirichlet one; accordingly, the latter is presented first.

7.A.1. Evaluation of the Dirichlet-Type Integral

In this section, the Dirichlet-type integral I_D, as reported in Equation (7.17), is evaluated: the expressions for $g(\mathbf{r}, \mathbf{r}')$ are provided by Equations (7.10); Equation (7.30) takes care of the surface-fields expansion, and the surface is described by means of Equation (7.1). Accordingly:

$$I^{\pm}_{D(1,2)} = -\frac{ik_{(1,2)}}{4\pi} \int_{-\infty}^{+\infty} dk_{(1,2)x} \frac{1}{k_{(1,2)z}} \exp\left[-i\left(k_{(1,2)x}x \pm k_{(1,2)z}z\right)\right]$$

$$\sum_{\substack{q_j=-\infty \\ j=0,\ldots,P-1}}^{+\infty} \alpha_{D,\mathbf{q}} \int_{-\infty}^{+\infty} dx' \exp\left[i\left(k_{(1,2)x} - k_{ix} - \tilde{\mathbf{q}} \cdot \boldsymbol{\kappa}\right)x'\right.$$

$$\left. \pm ik_{(1,2)z}z'\right], \quad (7.A.1)$$

where again the upper signs are obtained by letting $|z - z'| = z - z'$ in the exponential term of Equations (7.10), thus being referred to the fields in the half-space $z > z'_M$; conversely, the lower signs are obtained by letting $|z - z'| = -(z - z')$ in the exponential term of Equations (7.10), thus being referred to the fields in the half-space $z < z'_m$. Integration in the x' variable has been extended to the whole real axis: this is acceptable for surfaces large in terms of the incident electromagnetic wavelength.

The integral I_A in the x' variable, appearing in Equation (7.A.1),

$$I_A \triangleq \int_{-\infty}^{+\infty} dx' \exp\left[i\left(k_{(1,2)x} - k_{ix} - \tilde{\mathbf{q}} \cdot \boldsymbol{\kappa}\right)x' \pm ik_{(1,2)z}z'\right], \quad (7.A.2)$$

is now evaluated, where $z' = z'(x')$ and is provided by Equation (7.1). By using the Bessel identity in Equation (A.7.2), letting $a = k_{(1,2)z}BC_p\nu^{-Hp}$ and $\xi = \kappa_0\nu^p x' + \Phi_p$, it turns out that

$$\exp(\pm ik_{(1,2)z}z') = \prod_{p=0}^{P-1} \sum_{n=-\infty}^{\infty} J_n\left(\pm k_{(1,2)z}BC_p\nu^{-Hp}\right)$$

$$\exp\left[in\left(\kappa_0\nu^p x' + \Phi_p\right)\right]. \quad (7.A.3)$$

The Bessel identity in Equation (A.7.3), in conjunction with the distribution property in Equation (A.7.4), is now applied:

$$\exp(\pm ik_{(1,2)z}z') = \sum_{\substack{n_j=-\infty \\ j=0,\ldots,P-1}}^{+\infty} (\pm 1)^{m(\mathbf{n})} \exp(-i\tilde{\mathbf{n}} \cdot \boldsymbol{\Phi} - i\tilde{\mathbf{n}} \cdot \boldsymbol{\kappa} x')$$

$$\prod_{p=0}^{P-1} J_{n_p}\left(k_{(1,2)z} BC_p v^{-Hp}\right), \qquad (7.\text{A}.4)$$

wherein the function $m(\mathbf{n})$ is defined as

$$m(\mathbf{n}) \triangleq \sum_{p=0}^{P-1} n_p \qquad (7.\text{A}.5)$$

and

$$\tilde{\mathbf{n}} = [n_0, \ldots, n_{P-1}], \qquad (7.\text{A}.6)$$

$$\tilde{\boldsymbol{\Phi}} = [\Phi_0, \ldots, \Phi_{P-1}]. \qquad (7.\text{A}.7)$$

Substitution of Equation (7.A.4) in Equation (7.A.2) leads to

$$I_A = \sum_{\substack{n_j=-\infty \\ j=0,\ldots,P-1}}^{+\infty} (\pm 1)^{m(\mathbf{n})} \exp(-i\tilde{\mathbf{n}} \cdot \boldsymbol{\Phi}) \left\{ \int_{-\infty}^{+\infty} dx' \exp\left[-ik_{ix}x' - i(\tilde{\mathbf{n}} + \tilde{\mathbf{q}})\right.\right.$$

$$\left.\left. \cdot \boldsymbol{\kappa} x' + ik_{(1,2)x}x'\right] \right\} \prod_{p=0}^{P-1} J_{n_p}\left(k_{(1,2)z} BC_p v^{-Hp}\right)$$

$$= \sum_{\substack{n_j=-\infty \\ j=0,\ldots,P-1}}^{+\infty} (\pm 1)^{m(\mathbf{n})} \exp(-i\tilde{\mathbf{n}} \cdot \boldsymbol{\Phi})$$

$$\delta\left[k_{(1,2)x} - k_{ix} - (\tilde{\mathbf{n}} + \tilde{\mathbf{q}}) \cdot \boldsymbol{\kappa}\right] \prod_{p=0}^{P-1} J_{n_p}\left(k_{(1,2)z} BC_p v^{-Hp}\right). \qquad (7.\text{A}.8)$$

Equation (7.A.1) can be rearranged by taking advantage of Equation (7.A.8):

$$I_{D(1,2)}^{\pm} = -\frac{ik_{(1,2)}}{4\pi} \sum_{\substack{n_j=-\infty \\ j=0,\ldots,P-1}}^{+\infty} (\pm 1)^{m(\mathbf{n})} \exp(-i\tilde{\mathbf{n}} \cdot \boldsymbol{\Phi})$$

Appendix 7.A Evaluation of the Dirichlet- and Neumann-Type Integrals

$$\sum_{\substack{q_j=-\infty \\ j=0,\ldots,P-1}}^{+\infty} \alpha_{D,\mathbf{q}} \int_{-\infty}^{+\infty} dk_{(1,2)x} \cdot \frac{1}{k_{(1,2)z}} \exp\left[-i\left(k_{(1,2)x}x \pm k_{(1,2)z}z\right)\right]$$

$$\prod_{p=0}^{P-1} J_{n_p}\left(k_{(1,2)z}BC_p v^{-Hp}\right) \delta\left[k_{(1,2)x} - k_{ix} - (\tilde{\mathbf{n}} + \tilde{\mathbf{q}}) \cdot \boldsymbol{\kappa}\right]. \quad (7.A.9)$$

The spectral integral in Equation (7.A.9) can be evaluated by exploiting the sampling properties of the Dirac function:

$$I_{D(1,2)}^{\pm} = -\frac{ik_{(1,2)}}{4\pi} \sum_{\substack{n_j=-\infty \\ j=0,\ldots,P-1}}^{+\infty} (\pm 1)^{m(\mathbf{n})} \exp(-i\tilde{\mathbf{n}} \cdot \boldsymbol{\Phi})$$

$$\sum_{\substack{q_j=-\infty \\ j=0,\ldots,P-1}}^{+\infty} \alpha_{D,\mathbf{q}} \frac{1}{k_{(1,2)z\mathbf{l}}} \exp\left[-i\left(k_{x\mathbf{l}}x \pm k_{(1,2)z\mathbf{l}}z\right)\right]$$

$$\prod_{p=0}^{P-1} J_{n_p}\left(k_{(1,2)z\mathbf{l}}BC_p v^{-Hp}\right), \quad (7.A.10)$$

wherein the string $\mathbf{l} \in \mathbb{Z}^P$:

$$\mathbf{l} \triangleq \mathbf{n} + \mathbf{q}, \quad (7.A.11)$$

has been introduced, whose elements (in principle) span from $-\infty$ to $+\infty$. The wavenumber values, $k_{(1,2)x\mathbf{l}}$ and $k_{(1,2)z\mathbf{l}}$, in Equations (7.A.10) are sampled by the Dirac function, and are given by

$$k_{x\mathbf{l}} = k_{(1,2)x\mathbf{l}} = k_{ix} + \tilde{\mathbf{l}} \cdot \boldsymbol{\kappa}, \quad (7.A.12)$$

and

$$k_{(1,2)z\mathbf{l}} = \sqrt{k_{(1,2)}^2 - k_{x\mathbf{l}}^2} \quad (7.A.13)$$

upon considering Equations (7.11).

Equation (7.A.10) can be recast in a very compact form:

$$I_{D(1,2)}^{\pm} = \sum_{\substack{l_j=-\infty \\ j=0,\ldots,P-1}}^{+\infty} a_{D(1,2),\mathbf{l}}^{\pm} \exp\left(\mp i\mathbf{k}_{(1,2)\mathbf{l}}^{\pm} \cdot \mathbf{r}\right) \quad (7.A.14)$$

where

$$a^{\pm}_{D(1,2),\mathbf{l}} = -\frac{ik_{(1,2)}}{4\pi k_{(1,2)z\mathbf{l}}}(\pm 1)^{m(\mathbf{l})}\exp\left(-i\tilde{\mathbf{l}}\cdot\boldsymbol{\Phi}\right)\sum_{\substack{q_j=-\infty\\j=0,\ldots,P-1}}^{+\infty}\alpha_{D,\mathbf{q}}(\pm 1)^{m(-\mathbf{q})}$$

$$\exp(+i\tilde{\mathbf{q}}\cdot\boldsymbol{\Phi})\cdot\prod_{p=0}^{P-1}J_{l_p-q_p}\left(k_{(1,2)z\mathbf{l}}BC_p\nu^{-Hp}\right) \qquad (7.\mathrm{A}.15)$$

and

$$\mathbf{k}^{\pm}_{(1,2)\mathbf{l}} = k_{x\mathbf{l}}\hat{x} \pm k_{(1,2)z\mathbf{l}}\hat{z}. \qquad (7.\mathrm{A}.16)$$

A proper reading of Equation (7.A.14) is in order. The Dirichlet-type integral has been solved, and consists of a mode expansion: a P-infinity number of modes are present, each one specified by the choice of the P-indexes of the $\mathbf{l}$ string. Each mode is characterized by the wave vector $\mathbf{k}^{\pm}_{(1,2)\mathbf{l}}$ (Equation [7.A.16]). In the medium 1, the mode characterized by the string $\mathbf{l}$ propagates at an (possibly complex) angle $\theta_{1\mathbf{l}}$ with respect to the z-axis such that

$$k_{x\mathbf{l}} = k_1\sin\theta_{1\mathbf{l}} = k_1\sin\theta_i + \tilde{\mathbf{l}}\cdot\boldsymbol{\kappa}. \qquad (7.\mathrm{A}.17)$$

Similarly, in medium 2, the propagation angle for the mode $\mathbf{l}$ is given by

$$k_{x\mathbf{l}} = k_2\sin\theta_{2\mathbf{l}} = k_1\sin\theta_i + \tilde{\mathbf{l}}\cdot\boldsymbol{\kappa}. \qquad (7.\mathrm{A}.18)$$

Equations (7.A.17) and (7.A.18) are used also in the following section and in Section 7.5 to set the direction of propagation of each scattered or transmitted mode: they are usually referred to as *grating-modes equations*.

7.A.2. Evaluation of the Neumann-Type Integral

In this section, the Neumann-type integral I_N, as reported in Equation (7.18), is evaluated. The integrand can be exploded as

$$dS'\hat{\mathbf{n}}\cdot\nabla' g(\mathbf{r},\mathbf{r}') = dx'\left[-\frac{dz'(x')}{dx'}\frac{\partial}{\partial x'} + \frac{\partial}{\partial z'}\right]g(\mathbf{r},\mathbf{r}'); \qquad (7.\mathrm{A}.19)$$

the expressions for $g(\mathbf{r},\mathbf{r}')$ are provided by Equations (7.10); Equation (7.29) takes care of the surface fields in the integral; the surface is described by means of Equation (7.1). Accordingly:

$$I^{\pm}_{N(1,2)} = \frac{1}{4\pi}\int_{-\infty}^{+\infty}dk_{(1,2)x}\frac{1}{k_{(1,2)z}}\exp\left[-i\left(k_{(1,2)x}x \pm k_{(1,2)z}z\right)\right]$$

Appendix 7.A Evaluation of the Dirichlet- and Neumann-Type Integrals 235

$$\sum_{\substack{q_j=-\infty \\ j=0,\ldots,P-1}}^{+\infty} \alpha_{N,\mathbf{q}} \int_{-\infty}^{+\infty} dx' \left(k_{(1,2)x} \frac{dz'(x')}{dx'} \mp k_{(1,2)z} \right)$$

$$\exp\left[i\left(k_{(1,2)x} - k_{ix} - \tilde{\mathbf{q}} \cdot \boldsymbol{\kappa}\right)x' \pm ik_{(1,2)z}z'\right], \quad (7.\text{A}.20)$$

where again the upper sign corresponds to let $|z - z'| = z - z'$ in the exponential term of Equations (7.10), thus being referred to the fields in the half-plane $z > z'_M$; conversely, the lower sign corresponds to let $|z - z'| = -(z - z')$ in the exponential term of Equations (7.10), thus being referred to the fields in the half-plane $z < z'_m$. As in Section 7.A.1, integration in the x' variable has been extended to the whole real axis.

The integral I_B in the x' variable, appearing in Equation (7.A.20),

$$I_B \triangleq \int_{-\infty}^{+\infty} dx' \left(k_{(1,2)x} \frac{dz'(x')}{dx'} \mp k_{(1,2)z} \right)$$

$$\exp\left[i\left(k_{(1,2)x} - k_{ix} - \tilde{\mathbf{q}} \cdot \boldsymbol{\kappa}\right)x' \pm ik_{(1,2)z}z'\right]$$

$$= k_{(1,2)x} \int_{-\infty}^{+\infty} dx' \frac{dz'(x')}{dx'} \exp\left[i\left(k_{(1,2)x} - k_{ix} - \tilde{\mathbf{q}} \cdot \boldsymbol{\kappa}\right)x' \pm ik_{(1,2)z}z'\right]$$

$$\mp k_{(1,2)z} \int_{-\infty}^{+\infty} dx' \exp\left[i\left(k_{(1,2)x} - k_{ix} - \tilde{\mathbf{q}} \cdot \boldsymbol{\kappa}\right)x' \pm ik_{(1,2)z}z'\right]$$

$$= I_{B1} + I_{B2} \quad (7.\text{A}.21)$$

is now evaluated by splitting it into the two integrals I_{B1} and I_{B2}. Each one of the integrals I_{B1} and I_{B2} is now manipulated along two steps.

The relation

$$\frac{1}{k_{(1,2)z}} \frac{d}{dx'} \left\{ \exp\left[-i(k_{ix} + \tilde{\mathbf{q}} \cdot \boldsymbol{\kappa})x' \pm ik_{(1,2)z}z'\right] \right\}$$

$$= \pm i \frac{dz'}{dx'} \exp\left[-i(k_{ix} + \tilde{\mathbf{q}} \cdot \boldsymbol{\kappa})x' \pm ik_{(1,2)z}z'\right] - i \frac{(k_{ix} + \tilde{\mathbf{q}} \cdot \boldsymbol{\kappa})}{k_{(1,2)z}}$$

$$\exp\left[-i(k_{ix} + \tilde{\mathbf{q}} \cdot \boldsymbol{\kappa})x' \pm ik_{(1,2)z}z'\right] \quad (7.\text{A}.22)$$

is used to split I_{B1} into two new integrals I'_{B1} and I''_{B1} as follows:

$$I_{B1} \triangleq k_{(1,2)x} \int_{-\infty}^{+\infty} dx' \frac{dz'(x')}{dx'} \exp\left[i\left(k_{(1,2)x} - k_{ix} - \tilde{\mathbf{q}} \cdot \boldsymbol{\kappa}\right)x' \pm ik_{(1,2)z}z'\right]$$

$$= \mp i \frac{k_{(1,2)x}}{k_{(1,2)z}} \int_{-\infty}^{+\infty} dx' \exp\left(ik_{(1,2)x}x'\right)$$

$$\frac{d}{dx'}\left\{\exp\left[-i(k_{ix}+\tilde{\mathbf{q}}\cdot\boldsymbol{\kappa})x'\pm ik_{(1,2)z}z'\right]\right\}$$

$$\pm\frac{k_{(1,2)x}}{k_{(1,2)z}}(k_{ix}+\tilde{\mathbf{q}}\cdot\boldsymbol{\kappa})\int_{-\infty}^{+\infty}dx'\exp\left[i(k_{(1,2)x}-k_{ix}-\tilde{\mathbf{q}}\cdot\boldsymbol{\kappa})x'\pm ik_{(1,2)z}z'\right]$$

$$=I'_{B1}+I''_{B2}. \tag{7.A.23}$$

The first integral I'_{B1} in Equation (7.A.23) can be calculated by employing Equation (7.1) and Relation (7.A.4):

$$I'_{B1} \triangleq \mp i \frac{k_{(1,2)x}}{k_{(1,2)z}} \int_{-\infty}^{+\infty} dx'\, \exp(ik_{(1,2)x}x')$$

$$\frac{d}{dx'}\left\{\exp\left[-i(k_{ix}+\tilde{\mathbf{q}}\cdot\boldsymbol{\kappa})x' \pm ik_{(1,2)z}z'\right]\right\}$$

$$= \mp i \frac{k_{(1,2)x}}{k_{(1,2)z}} \sum_{\substack{n_j=-\infty \\ j=0,\ldots,P-1}}^{+\infty} (\pm 1)^{m(\mathbf{n})} \exp(-i\tilde{\mathbf{n}}\cdot\boldsymbol{\Phi}) \prod_{p=0}^{P-1} J_{n_p}\left(k_{(1,2)z}BC_p \nu^{-Hp}\right)$$

$$\int_{-\infty}^{+\infty} dx'\, \exp\left\{i\left[k_{(1,2)x}-k_{ix}-(\tilde{\mathbf{q}}+\tilde{\mathbf{n}})\cdot\boldsymbol{\kappa}\right]x'\right\}$$

$$\left[-i(k_{ix}+(\tilde{\mathbf{q}}+\tilde{\mathbf{n}})\cdot\boldsymbol{\kappa})\right]. \tag{7.A.24}$$

Proceeding as in Section 7.A.1, the final result for I'_{B1},

$$I'_{B1} = \mp \sum_{\substack{n_j=-\infty \\ j=0,\ldots,P-1}}^{+\infty} (\pm 1)^{m(\mathbf{n})} \exp(-i\tilde{\mathbf{n}}\cdot\boldsymbol{\Phi}) \frac{k_{(1,2)x}}{k_{(1,2)z}} [k_{ix}+(\tilde{\mathbf{q}}+\tilde{\mathbf{n}})\cdot\boldsymbol{\kappa}]$$

$$\delta\left[k_{(1,2)x}-k_{ix}-(\tilde{\mathbf{q}}+\tilde{\mathbf{n}})\boldsymbol{\kappa}\right] \prod_{p=0}^{P-1} J_{n_p}\left(k_{(1,2)z}BC_p \nu^{-Hp}\right), \tag{7.A.25}$$

is obtained.

The second integral I''_{B1} in Equation (7.A.23) is calculated along the same lines:

$$I''_{B1} = \pm \frac{k_{(1,2)x}}{k_{(1,2)z}}(k_{ix}+\tilde{\mathbf{q}}\cdot\boldsymbol{\kappa})$$

$$\int_{-\infty}^{+\infty} dx'\, \exp\left[i\left(k_{(1,2)x}-k_{ix}-\tilde{\mathbf{q}}\cdot\boldsymbol{\kappa}\right)x' \pm ik_{(1,2)z}z'\right]$$

Appendix 7.A Evaluation of the Dirichlet- and Neumann-Type Integrals   237

$$= \pm \frac{k_{(1,2)x}}{k_{(1,2)z}} (k_{ix} + \tilde{\mathbf{q}} \cdot \boldsymbol{\kappa}) I_A. \qquad (7.A.26)$$

The integral I_{B2} in Equation (7.A.21) is of the same form of the integral defined in Equation (7.A.2). Then it turns out that

$$I_{B2} \overset{\triangle}{=} \mp k_{(1,2)z} \int_{-\infty}^{+\infty} dx' \exp\left[i(k_{(1,2)x} - k_{ix} - \tilde{\mathbf{q}} \cdot \boldsymbol{\kappa})x' \pm ik_{(1,2)z}z'\right]$$

$$= \mp k_{(1,2)z} I_A. \qquad (7.A.27)$$

In conclusion, it is possible to express the integral I_N by substituting first Equations (7.A.25) and (7.A.26) in Equation (7.A.23), then Equations (7.A.23) and (7.A.27) in Equation (7.A.21), and finally accounting for Equation (7.A.8). It turns out that:

$$I_{N(1,2)}^\pm = \sum_{\substack{l_j = -\infty \\ j=0,\dots,P-1}}^{+\infty} a_{N(1,2),\mathbf{l}}^\pm \exp\left(\mp i \mathbf{k}_{(1,2)\mathbf{l}}^\pm \cdot \mathbf{r}\right), \qquad (7.A.28)$$

where

$$a_{N(1,2),\mathbf{l}}^\pm = -\frac{k_{(1,2)}^2 - k_{x\mathbf{l}} k_{x\mathbf{q}}}{4\pi k_{(1,2)z\mathbf{l}}^2} (\pm 1)^{m(\mathbf{l})} \exp\left(-i\tilde{\mathbf{l}} \cdot \boldsymbol{\Phi}\right)$$

$$\sum_{\substack{q_j = -\infty \\ j=0,\dots,P-1}}^{+\infty} \alpha_{N,\mathbf{q}} (\pm 1)^{m(\mathbf{q})} \exp(i\tilde{\mathbf{q}} \cdot \boldsymbol{\Phi})$$

$$\prod_{p=0}^{P-1} J_{l_p - q_p}\left(k_{(1,2)z\mathbf{l}} BC_p v^{-Hp}\right). \qquad (7.A.29)$$

Equation (7.A.28) represents a superposition of modes, similar to Equation (7.A.14). Accordingly, the same considerations at the end of Section 7.A.1 do apply as far as number of modes and their direction of propagation are concerned; more specifically, the same grating-modes Equations (7.A.17) and (7.A.18) hold.

CHAPTER 8

Scattering from Fractional Brownian Surfaces: Small-Perturbation Method

8.1. Introduction and Chapter Outline

In this chapter the Small Perturbation Method (SPM) is employed to solve the scattered power density problem whenever the fractional Brownian Model (fBm) is used to describe the surface. The procedure presented in this chapter is based on the EBCM formulation provided in Chapter 4. The general case of dielectric surfaces with a two-dimensional topological dimension is considered. However, the SPM closed-form solution for the scattered power density is obtained only under appropriate roughness regimes; more specifically, in the SPM the surface fields and the surface profile are expanded in power series and the solution to the scattering problem is iteratively obtained for each series term; then, the total scattered power density is generated by addition of the obtained different order solutions. Zero- and first-order solutions are presented in detail in this chapter, whereas only the rationale for higher order solutions is provided instead. It is shown that the zero-order solution generates the scattered field coherent component, whereas the first-order solution gives rise to a surface-dependent incoherent contribution. The key point of the procedure is the property of the fBm to be represented by the simple, well defined, spectral behavior characterized by two parameters, that has been presented under Chapter 3.

The chapter is organized as follows. Rationale to obtain the SPM solution is described in Section 8.2. The way to express the EBCM equations in the transformed domain is presented in Section 8.3. Set up of the SPM solution is provided in Section 8.4. An appropriate coordinate system, which includes the normal to the surface mean plane, is introduced in Section 8.5: it allows simplifying the notation in the small roughness regimes. Then, the zero-order SPM solution is derived in Section 8.6. Results corresponding to the first-order SPM solution are presented in Section 8.7. Limits of validity for the SPM are evaluated in Section 8.8. Numerical results and scattering diagrams are presented and discussed in Section 8.9. References and further readings are reported in Section 8.10.

8.2. Rationale of the SPM Solution

As stated in Chapter 4, for handling the scattering problem depicted in Figure 8.1, the SPM makes use of the Integral Equations (4.61)

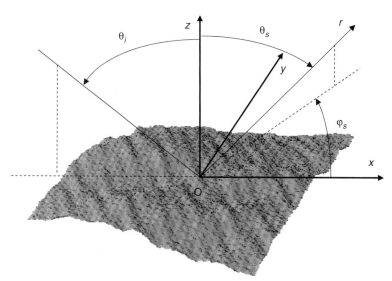

FIGURE 8.1 Geometry of the scattering problem. Cartesian and polar reference systems relevant to the scattering surface are depicted.

8.2. Rationale of the SPM Solution

through (4.64), recalled hereafter for convenience:

$$\mathbf{E}^{(i)}(\mathbf{r}) = \frac{1}{8\pi^2} \int_{-\infty}^{+\infty} d\mathbf{k}_\perp \exp[-i(\mathbf{k}_\perp - k_{1z}\hat{\mathbf{z}}) \cdot \mathbf{r}] \frac{k_1}{k_{1z}}$$
$$\int_{-\infty}^{+\infty} d\mathbf{r}'_\perp \exp[i(\mathbf{k}_\perp - k_{1z}\hat{\mathbf{z}}) \cdot \mathbf{r}'] \left\{ \left[\hat{\mathbf{v}}_1^- \hat{\mathbf{v}}_1^- + \hat{\mathbf{h}}_1^- \hat{\mathbf{h}}_1^- \right] \cdot \mathbf{a}(\mathbf{r}'_\perp) \right.$$
$$\left. + \left[\hat{\mathbf{h}}_1^- \hat{\mathbf{v}}_1^- - \hat{\mathbf{v}}_1^- \hat{\mathbf{h}}_1^- \right] \cdot \mathbf{b}(\mathbf{r}'_\perp) \right\}, \quad \forall z < z_m, \tag{8.1}$$

$$0 = \frac{1}{8\pi^2} \int_{-\infty}^{+\infty} d\mathbf{k}_\perp \exp[-i(\mathbf{k}_\perp + k_{2z}\hat{\mathbf{z}}) \cdot \mathbf{r}] \frac{k_2}{k_{2z}}$$
$$\int_{-\infty}^{+\infty} d\mathbf{r}'_\perp \exp[i(\mathbf{k}_\perp + k_{2z}\hat{\mathbf{z}}) \cdot \mathbf{r}'] \left\{ \frac{k_1}{k_2} \left[\hat{\mathbf{v}}_2^+ \hat{\mathbf{v}}_2^+ + \hat{\mathbf{h}}_2^+ \hat{\mathbf{h}}_2^+ \right] \cdot \mathbf{a}(\mathbf{r}'_\perp) \right.$$
$$\left. + \left[\hat{\mathbf{h}}_2^+ \hat{\mathbf{v}}_2^+ - \hat{\mathbf{v}}_2^+ \hat{\mathbf{h}}_2^+ \right] \cdot \mathbf{b}(\mathbf{r}'_\perp) \right\}, \quad \forall z > z_M, \tag{8.2}$$

$$\mathbf{E}^{(s)}(\mathbf{r}) = \mathbf{E}(\mathbf{r}) - \mathbf{E}^{(i)}(\mathbf{r})$$
$$= -\frac{1}{8\pi^2} \int_{-\infty}^{+\infty} d\mathbf{k}_\perp \exp[-i(\mathbf{k}_\perp + k_{1z}\hat{\mathbf{z}}) \cdot \mathbf{r}] \frac{k_1}{k_{1z}}$$
$$\int_{-\infty}^{+\infty} d\mathbf{r}'_\perp \exp[i(\mathbf{k}_\perp + k_{1z}\hat{\mathbf{z}}) \cdot \mathbf{r}'] \left\{ \left[\hat{\mathbf{v}}_1^+ \hat{\mathbf{v}}_1^+ + \hat{\mathbf{h}}_1^+ \hat{\mathbf{h}}_1^+ \right] \cdot \mathbf{a}(\mathbf{r}'_\perp) \right.$$
$$\left. + \left[\hat{\mathbf{h}}_1^+ \hat{\mathbf{v}}_1^+ - \hat{\mathbf{v}}_1^+ \hat{\mathbf{h}}_1^+ \right] \cdot \mathbf{b}(\mathbf{r}'_\perp) \right\}, \quad \forall z > z_M, \tag{8.3}$$

$$\mathbf{E}^{(t)}(\mathbf{r}) = \frac{1}{8\pi^2} \int_{-\infty}^{+\infty} d\mathbf{k}_\perp \exp[-i(\mathbf{k}_\perp - k_{2z}\hat{\mathbf{z}}) \cdot \mathbf{r}] \frac{k_2}{k_{2z}}$$
$$\int_{-\infty}^{+\infty} d\mathbf{r}'_\perp \exp[i(\mathbf{k}_\perp - k_{2z}\hat{\mathbf{z}}) \cdot \mathbf{r}'] \left\{ \frac{k_1}{k_2} \left[\hat{\mathbf{v}}_2^- \hat{\mathbf{v}}_2^- + \hat{\mathbf{h}}_2^- \hat{\mathbf{h}}_2^- \right] \cdot \mathbf{a}(\mathbf{r}'_\perp) \right.$$
$$\left. + \left[\hat{\mathbf{h}}_2^- \hat{\mathbf{v}}_2^- - \hat{\mathbf{v}}_2^- \hat{\mathbf{h}}_2^- \right] \cdot \mathbf{b}(\mathbf{r}'_\perp) \right\}, \quad \forall z < z_m, \tag{8.4}$$

that have been obtained upon use of the Greens' functions evaluated in the spectral domain. The scattered and transmitted electromagnetic fields at the generic point **r** in the upper or lower medium are formally represented in the left-hand side member of the integral Equations (8.1) and (8.2); they are obtained, as indicated in the corresponding right-hand side member, by considering the contributions provided by the unknown surface fields $\mathbf{a}(\mathbf{r}'_\perp)$ and $\mathbf{b}(\mathbf{r}'_\perp)$, that are distributed over the scattering surface whose generic point position is indicated by the vector $\mathbf{r}'_\perp$. These surface fields are solutions of Equations (8.1) and (8.2).

In Equations (8.1) through (8.4), the surface profile is still included in the exponential term $\mathbf{r}'$ inside the diffraction integral. Hence, the scattered field is dependent on the surface geometrical shape. For fBm surfaces, this is described by a regular stochastic process: accordingly, the scattered field is evaluated as a regular stochastic process, too.

Before proceeding further, manipulation of Equations (8.1) through (8.4) is convenient. This requires a number of operations, likely to be uneasy to follow due to the heavy notations appearing in the equations themselves. For this reason the rationale of the procedure to be implemented is presented first; then, the analytical details are provided.

First of all, the exponential functions, $\exp[\pm ik_{(1,2)z}z'(\mathbf{r}'_\perp)]$ appearing in the equations are expanded in their Taylor power series. The reason is to move in the direction of the SPM solution: when the argument of the exponentials is small, the power series terms becomes increasingly negligible as their order increases. In this hypothesis, it is obviously expected that the final solution could retain only a finite number of the expansion terms: the n-order solution is referred to when terms up to the n-th order are retained.

The second step is suggested by examination of the equations, each one consisting of a two-fold integral, the outer in the spectral and the inner in the space domain. Both integrals contain a FT-type kernel. In particular, products of powers of the surface profile function and the surface fields, i.e., the unknowns to be determined, appear in the inner integral: accordingly, this integral, that contains a Fourier kernel, is transformed in the convolution between the FTs of the two factors appearing in the above mentioned products, i.e., the FT of the powers of the surface profile and the FT of the unknowns. This formulation, with the convolution integral operating on the spectral domain variables, simplifies the computational procedure as it is described in the sections devoted to the analytical details.

The successive step is the elimination of the outer FT-type integral appearing at the right-hand side of the equations. This task is easily accomplished by enforcing a similar FT at the left-hand side of the equations, and then equating the integrands. The final result is a set of equations with only a convolution-type integral at their right-hand side.

Then, the spirit of the SPM solution is fully exploited, as preparation to the final step: the expansion of the exponential factors inside the integrals

8.3. Extended Boundary Condition Method in the Transformed Domain

suggests a similar expansion for the unknowns. Accordingly, these are expanded in a series of unknown terms, postulating that each one of them is of the same order of the corresponding surface profile one.

The final step is implemented by inserting the unknown expansions in the integrals, performing all the mutual products, classifying the latter according to their resulting order and enforcing the equations at each order level. It is easily anticipated that the resulting equations at order n would contain lower level order terms of the unknowns: the solution proceeds in recursive fashion, so that computation of the n-th order term of the unknowns requires knowledge of all previous order terms of the unknowns themselves. Additional details must be accounted for: for instance, the z-component of the surface profile is of higher order compared to the transverse component, if the surface slope is taken small (small slope approximation). But these details do not impair the presented procedure that is detailed in the following.

8.3. Extended Boundary Condition Method in the Transformed Domain

SPM is employed to solve Equations (8.1) and (8.2) and compute Equations (8.3) and (8.4). This method requires that the exponential terms involving the surface profile in the integrals of Equations (8.1) through (8.4) are expanded in power series of $k_{(1,2)z}z'$:

$$\exp[\pm ik_{(1,2)z}z'(\mathbf{r}'_\perp)] = \sum_{m=0}^{\infty} \frac{[\pm ik_{(1,2)z}z'(\mathbf{r}'_\perp)]^m}{m!}. \tag{8.5}$$

Equation (8.5) is substituted in Equations (8.1) through (8.4), thus providing:

$$\begin{aligned}
\mathbf{E}^{(i)}(\mathbf{r}) = & \frac{1}{8\pi^2} \int_{-\infty}^{+\infty} d\mathbf{k}_\perp \exp[-i(\mathbf{k}_\perp - k_{1z}\hat{\mathbf{z}}) \cdot \mathbf{r}] \frac{k_1}{k_{1z}} \\
& \int_{-\infty}^{+\infty} d\mathbf{r}'_\perp \exp[i\mathbf{k}_\perp \cdot \mathbf{r}'] \left\{ \left[\hat{\mathbf{v}}_1^- \hat{\mathbf{v}}_1^- + \hat{\mathbf{h}}_1^- \hat{\mathbf{h}}_1^- \right] \cdot \mathbf{a}(\mathbf{r}'_\perp) \right. \\
& \left. + \left[\hat{\mathbf{h}}_1^- \hat{\mathbf{v}}_1^- - \hat{\mathbf{v}}_1^- \hat{\mathbf{h}}_1^- \right] \cdot \mathbf{b}(\mathbf{r}'_\perp) \right\} \sum_{m=0}^{\infty} \frac{[-ik_{1z}z'(\mathbf{r}'_\perp)]^m}{m!}, \quad \forall z < z_m,
\end{aligned} \tag{8.6}$$

244   8 ◊ Scattering from fBm Surfaces: SPM Solution

$$0 = \frac{1}{8\pi^2} \int_{-\infty}^{+\infty} d\mathbf{k}_\perp \exp[-i(\mathbf{k}_\perp + k_{2z}\hat{\mathbf{z}}) \cdot \mathbf{r}] \frac{k_2}{k_{2z}}$$

$$\int_{-\infty}^{+\infty} d\mathbf{r}'_\perp \exp[i\mathbf{k}_\perp \cdot \mathbf{r}'] \left\{ \frac{k_1}{k_2} \left[\hat{\mathbf{v}}_2^+ \hat{\mathbf{v}}_2^+ + \hat{\mathbf{h}}_2^+ \hat{\mathbf{h}}_2^+ \right] \cdot \mathbf{a}(\mathbf{r}'_\perp) \right.$$

$$\left. + \left[\hat{\mathbf{h}}_2^+ \hat{\mathbf{v}}_2^+ - \hat{\mathbf{v}}_2^+ \hat{\mathbf{h}}_2^+ \right] \cdot \mathbf{b}(\mathbf{r}'_\perp) \right\} \sum_{m=0}^{\infty} \frac{[ik_{2z}z'(\mathbf{r}'_\perp)]^m}{m!}, \quad \forall z > z_M,$$

(8.7)

$$\mathbf{E}^{(s)}(\mathbf{r}) = -\frac{1}{8\pi^2} \int_{-\infty}^{+\infty} d\mathbf{k}_\perp \exp[-i(\mathbf{k}_\perp + k_{1z}\hat{\mathbf{z}}) \cdot \mathbf{r}] \frac{k_1}{k_{1z}}$$

$$\int_{-\infty}^{+\infty} d\mathbf{r}'_\perp \exp[i\mathbf{k}_\perp \cdot \mathbf{r}'] \left\{ \left[\hat{\mathbf{v}}_1^+ \hat{\mathbf{v}}_1^+ + \hat{\mathbf{h}}_1^+ \hat{\mathbf{h}}_1^+ \right] \cdot \mathbf{a}(\mathbf{r}'_\perp) \right.$$

$$\left. + \left[\hat{\mathbf{h}}_1^+ \hat{\mathbf{v}}_1^+ - \hat{\mathbf{v}}_1^+ \hat{\mathbf{h}}_1^+ \right] \cdot \mathbf{b}(\mathbf{r}'_\perp) \right\} \sum_{m=0}^{\infty} \frac{[ik_{1z}z'(\mathbf{r}'_\perp)]^m}{m!}, \quad \forall z > z_M,$$

(8.8)

$$\mathbf{E}^{(t)}(\mathbf{r}) = \frac{1}{8\pi^2} \int_{-\infty}^{+\infty} d\mathbf{k}_\perp \exp[-i(\mathbf{k}_\perp - k_{2z}\hat{\mathbf{z}}) \cdot \mathbf{r}] \frac{k_2}{k_{2z}}$$

$$\int_{-\infty}^{+\infty} d\mathbf{r}'_\perp \exp[i\mathbf{k}_\perp \cdot \mathbf{r}'] \left\{ \frac{k_1}{k_2} \left[\hat{\mathbf{v}}_2^- \hat{\mathbf{v}}_2^- + \hat{\mathbf{h}}_2^- \hat{\mathbf{h}}_2^- \right] \cdot \mathbf{a}(\mathbf{r}'_\perp) \right.$$

$$\left. + \left[\hat{\mathbf{h}}_2^- \hat{\mathbf{v}}_2^- - \hat{\mathbf{v}}_2^- \hat{\mathbf{h}}_2^- \right] \cdot \mathbf{b}(\mathbf{r}'_\perp) \right\} \sum_{m=0}^{\infty} \frac{[-ik_{2z}z'(\mathbf{r}'_\perp)]^m}{m!}, \quad \forall z < z_m.$$

(8.9)

In Equations (8.6) through (8.9) the integrals in the variable $\mathbf{r}'_\perp$ are operated over the scattering surface, which is assumed here of infinite extent for the sake of simplicity. Use of Equation (8.5) leads to read these integrals as Fourier Transforms of the products of the unknown surface fields $\mathbf{a}$ and $\mathbf{b}$ with the powers of the surface profile $z'(\mathbf{r}'_\perp)$. A more compact expression for the Equation (8.6) through (8.9) can then be obtained by introducing $\mathbf{A}(\mathbf{k}_\perp)$ and $\mathbf{B}(\mathbf{k}_\perp)$, the Fourier domain counterparts of the vector $\mathbf{a}(\mathbf{r}'_\perp)$

8.3. Extended Boundary Condition Method in the Transformed Domain

and $\mathbf{b}(\mathbf{r}'_\perp)$ unknowns functions:

$$\mathbf{A}(\mathbf{k}_\perp) = \frac{1}{(2\pi)^2} \int_{-\infty}^{\infty} d\mathbf{r}'_\perp \mathbf{a}(\mathbf{r}'_\perp) \exp(i\mathbf{k}_\perp \cdot \mathbf{r}'_\perp)$$

$$\mathbf{B}(\mathbf{k}_\perp) = \frac{1}{(2\pi)^2} \int_{-\infty}^{\infty} d\mathbf{r}'_\perp \mathbf{b}(\mathbf{r}'_\perp) \exp(i\mathbf{k}_\perp \cdot \mathbf{r}'_\perp)$$

(8.10)

as well as of $Z'^{(m)}(\mathbf{k}_\perp)$ the generalized Fourier Transform (FT) of $z'(\mathbf{r}'_\perp)$, i.e., the FT of the m-power, $z'^m(\mathbf{r}'_\perp)$, of the scattering surface shape $z'(\mathbf{r}'_\perp)$:

$$Z'^{(m)}(\mathbf{k}_\perp) = \frac{1}{(2\pi)^2} \int_{-\infty}^{\infty} d\mathbf{r}_\perp \exp(i\mathbf{k}_\perp \cdot \mathbf{r}_\perp) z'^m(\mathbf{r}_\perp). \quad (8.11)$$

Substitution of Equations (8.10) and (8.11) in Equations (8.6) through (8.9) and use of the Borel theorem generates the new equations in the FTs of the surface unknowns, $\mathbf{A}(\mathbf{k}_\perp)$, $\mathbf{B}(\mathbf{k}_\perp)$:

$$\mathbf{E}^{(i)}(\mathbf{r}) = \frac{1}{2} \int_{-\infty}^{\infty} d\mathbf{k}_\perp \exp(-i\mathbf{k}_\perp \cdot \mathbf{r}_\perp + ik_{1z}z)$$
$$\frac{k_1}{k_{1z}} \left\{ \left[\hat{\mathbf{v}}_1^- \hat{\mathbf{v}}_1^- + \hat{\mathbf{h}}_1^- \hat{\mathbf{h}}_1^- \right] \right.$$
$$\cdot \left[\sum_{m=0}^{\infty} (-ik_{1z})^m \int d\mathbf{k}'_\perp \mathbf{A}(\mathbf{k}'_\perp) Z'^{(m)}(\mathbf{k}_\perp - \mathbf{k}'_\perp) \right]$$
$$+ \left[\hat{\mathbf{h}}_1^- \hat{\mathbf{v}}_1^- - \hat{\mathbf{v}}_1^- \hat{\mathbf{h}}_1^- \right] \cdot \left[\sum_{m=0}^{\infty} (-ik_{1z})^m \right.$$
$$\left. \left. \int d\mathbf{k}'_\perp \mathbf{B}(\mathbf{k}'_\perp) Z'^{(m)}(\mathbf{k}_\perp - \mathbf{k}'_\perp) \right] \right\}, \quad \forall z < z_m, \quad (8.12)$$

$$0 = \frac{1}{2} \int_{-\infty}^{\infty} d\mathbf{k}_\perp \exp(-i\mathbf{k}_\perp \cdot \mathbf{r}_\perp - ik_{2z}z) \frac{k_2}{k_{2z}} \left\{ \frac{k_1}{k_2} \left[\hat{\mathbf{v}}_2^+ \hat{\mathbf{v}}_2^+ + \hat{\mathbf{h}}_2^+ \hat{\mathbf{h}}_2^+ \right] \right.$$
$$\cdot \left[\sum_{m=0}^{\infty} (ik_{2z})^m \int d\mathbf{k}'_\perp \mathbf{A}(\mathbf{k}'_\perp) Z'^{(m)}(\mathbf{k}_\perp - \mathbf{k}'_\perp) \right]$$
$$+ \left[\hat{\mathbf{h}}_2^+ \hat{\mathbf{v}}_2^+ - \hat{\mathbf{v}}_2^+ \hat{\mathbf{h}}_2^+ \right] \cdot \left[\sum_{m=0}^{\infty} (ik_{2z})^m \right.$$
$$\left. \left. \int d\mathbf{k}'_\perp \mathbf{B}(\mathbf{k}'_\perp) Z'^{(m)}(\mathbf{k}_\perp - \mathbf{k}'_\perp) \right] \right\}, \quad \forall z > z_M, \quad (8.13)$$

$$\mathbf{E}^{(s)}(\mathbf{r}) = -\frac{1}{2}\int_{-\infty}^{\infty} d\mathbf{k}_{\perp}\exp(-i\mathbf{k}_{\perp}\cdot\mathbf{r}_{\perp} - ik_{1z}z)\frac{k_1}{k_{1z}}\left\{\left[\hat{\mathbf{v}}_1^+\hat{\mathbf{v}}_1^+ + \hat{\mathbf{h}}_1^+\hat{\mathbf{h}}_1^+\right]\right.$$
$$\cdot\left[\sum_{m=0}^{\infty}(ik_{1z})^m\int d\mathbf{k}'_{\perp}\mathbf{A}(\mathbf{k}'_{\perp})Z'^{(m)}(\mathbf{k}_{\perp}-\mathbf{k}'_{\perp})\right]$$
$$+\left[\hat{\mathbf{h}}_1^+\hat{\mathbf{v}}_1^+ - \hat{\mathbf{v}}_1^+\hat{\mathbf{h}}_1^+\right]\cdot\left[\sum_{m=0}^{\infty}(ik_{1z})^m\right.$$
$$\left.\left.\int d\mathbf{k}'_{\perp}\mathbf{B}(\mathbf{k}'_{\perp})Z'^{(m)}(\mathbf{k}_{\perp}-\mathbf{k}'_{\perp})\right]\right\}, \quad \forall z > z_M, \qquad (8.14)$$

$$\mathbf{E}^{(t)}(\mathbf{r}) = \frac{1}{2}\int_{-\infty}^{\infty} d\mathbf{k}_{\perp}\exp(-i\mathbf{k}_{\perp}\cdot\mathbf{r}_{\perp} + ik_{2z}\hat{\mathbf{z}}\cdot\mathbf{r})\frac{k_2}{k_{2z}}\left\{\frac{k_1}{k_2}\left[\hat{\mathbf{v}}_2^-\hat{\mathbf{v}}_2^- + \hat{\mathbf{h}}_2^-\hat{\mathbf{h}}_2^-\right]\right.$$
$$\cdot\left[\sum_{m=0}^{\infty}(-ik_{2z})^m\int d\mathbf{k}'_{\perp}\mathbf{A}(\mathbf{k}'_{\perp})Z'^{(m)}(\mathbf{k}_{\perp}-\mathbf{k}'_{\perp})\right]$$
$$+\left[\hat{\mathbf{h}}_2^-\hat{\mathbf{v}}_2^- - \hat{\mathbf{v}}_2^-\hat{\mathbf{h}}_2^-\right]\cdot\left[\sum_{m=0}^{\infty}(-ik_{2z})^m\right.$$
$$\left.\left.\int d\mathbf{k}'_{\perp}\mathbf{B}(\mathbf{k}'_{\perp})Z'^{(m)}(\mathbf{k}_{\perp}-\mathbf{k}'_{\perp})\right]\right\}, \quad \forall z < z_m. \qquad (8.15)$$

Equations (8.12) and (8.13) must be solved to get the FT of the surface current expressions in the spectral domain. Then, Equations (8.14) and (8.15) can be employed to get the scattered and transmitted field, respectively.

Solution of Equations (8.12) and (8.13) is in order. Their left-hand side members, representing the incident and a null fields, can be expanded in plane waves. For a single plane wave incidence, see Figure 8.1, the incident field is given by:

$$\mathbf{E}^{(i)}(\mathbf{r}) = E_i\exp(-i\mathbf{k}_{i\perp}\cdot\mathbf{r}_{\perp} + ik_{iz}z)\hat{\mathbf{e}}_i$$
$$= \int_{-\infty}^{\infty} d\mathbf{k}_{\perp}E_i\exp[-i\mathbf{k}_{\perp}\cdot\mathbf{r}_{\perp} + ik_{1z}z]\hat{\mathbf{e}}_i\delta(\mathbf{k}_{\perp}-\mathbf{k}_{i\perp}) \qquad (8.16)$$

where E_i is the amplitude of the incident field, and $\mathbf{k}_i = k_{ix}\hat{\mathbf{x}} + k_{iy}\hat{\mathbf{y}} + k_{iz}\hat{\mathbf{z}} = \mathbf{k}_{i\perp} + k_{iz}\hat{\mathbf{z}}$ is the incident wavenumber vector. Extension to more than a single plane wave can be accomplished by means of plane waves expansion.

8.3. Extended Boundary Condition Method in the Transformed Domain

Equations (8.12) and (8.13) are now considered. The expression of the incident field in Equation (8.16) is substituted in Equations (8.12) and (8.13):

$$\int_{-\infty}^{\infty} d\mathbf{k}_\perp E_i \exp[-i\mathbf{k}_\perp \cdot \mathbf{r}_\perp + ik_{1z}\hat{\mathbf{z}} \cdot \mathbf{r}]\hat{\mathbf{e}}_i \delta(\mathbf{k}_\perp - \mathbf{k}_{i\perp})$$

$$= \frac{1}{2}\int_{-\infty}^{\infty} d\mathbf{k}_\perp \exp(-i\mathbf{k}_\perp \cdot \mathbf{r}_\perp + ik_{1z}z) \frac{k_1}{k_{1z}} \left\{ \left[\hat{\mathbf{v}}_1^- \hat{\mathbf{v}}_1^- + \hat{\mathbf{h}}_1^- \hat{\mathbf{h}}_1^- \right] \right.$$

$$\cdot \left[\sum_{m=0}^{\infty}(-ik_{1z})^m \int d\mathbf{k}'_\perp \mathbf{A}(\mathbf{k}'_\perp) Z'^{(m)}(\mathbf{k}_\perp - \mathbf{k}'_\perp)\right] + \left[\hat{\mathbf{h}}_1^- \hat{\mathbf{v}}_1^- - \hat{\mathbf{v}}_1^- \hat{\mathbf{h}}_1^- \right]$$

$$\left. \cdot \left[\sum_{m=0}^{\infty}(-ik_{1z})^m \int d\mathbf{k}'_\perp \mathbf{B}(\mathbf{k}'_\perp) Z'^{(m)}(\mathbf{k}_\perp - \mathbf{k}'_\perp)\right] \right\}, \quad \forall z < z_m,$$

(8.17)

$$0 = \frac{1}{2}\int_{-\infty}^{\infty} d\mathbf{k}_\perp \exp(-i\mathbf{k}_\perp \cdot \mathbf{r}_\perp - ik_{2z}z) \frac{k_2}{k_{2z}} \left\{ \frac{k_1}{k_2}\left[\hat{\mathbf{v}}_2^+ \hat{\mathbf{v}}_2^+ + \hat{\mathbf{h}}_2^+ \hat{\mathbf{h}}_2^+ \right] \right.$$

$$\cdot \left[\sum_{m=0}^{\infty}(ik_{2z})^m \int d\mathbf{k}'_\perp \mathbf{A}(\mathbf{k}'_\perp) Z'^{(m)}(\mathbf{k}_\perp - \mathbf{k}'_\perp)\right] + \left[\hat{\mathbf{h}}_2^+ \hat{\mathbf{v}}_2^+ - \hat{\mathbf{v}}_2^+ \hat{\mathbf{h}}_2^+ \right]$$

$$\left. \cdot \left[\sum_{m=0}^{\infty}(ik_{2z})^m \int d\mathbf{k}'_\perp \mathbf{B}(\mathbf{k}'_\perp) Z'^{(m)}(\mathbf{k}_\perp - \mathbf{k}'_\perp)\right] \right\}, \quad \forall z > z_M \quad (8.18)$$

Equations (8.17) and (8.18) consist of identities between FTs in the variable $\mathbf{k}_\perp$ (the zero value at the left-hand side of Equation (8.18) can be read as the plane wave expansion of a null field). Accordingly, their inverse FT (which is here equivalent to equate the integrands) leads to:

$$E_i \hat{\mathbf{e}}_i \delta(\mathbf{k}_\perp - \mathbf{k}_{i\perp}) = \frac{1}{2}\frac{k_1}{k_{1z}} \left\{ \left[\hat{\mathbf{v}}_1^- \hat{\mathbf{v}}_1^- + \hat{\mathbf{h}}_1^- \hat{\mathbf{h}}_1^- \right] \right.$$

$$\cdot \left[\sum_{m=0}^{\infty}(-ik_{1z})^m \int d\mathbf{k}'_\perp \mathbf{A}(\mathbf{k}'_\perp) Z'^{(m)}(\mathbf{k}_\perp - \mathbf{k}'_\perp)\right]$$

$$+ \left[\hat{\mathbf{h}}_1^- \hat{\mathbf{v}}_1^- - \hat{\mathbf{v}}_1^- \hat{\mathbf{h}}_1^- \right] \cdot \left[\sum_{m=0}^{\infty}(-ik_{1z})^m \right.$$

$$\left. \left. \int d\mathbf{k}'_\perp \mathbf{B}(\mathbf{k}'_\perp) Z'^{(m)}(\mathbf{k}_\perp - \mathbf{k}'_\perp)\right] \right\}, \quad \forall z < z_m, \quad (8.19)$$

$$0 = \frac{1}{2}\frac{k_2}{k_{2z}}\left\{\frac{k_1}{k_2}\left[\hat{\mathbf{v}}_2^+\hat{\mathbf{v}}_2^+ + \hat{\mathbf{h}}_2^+\hat{\mathbf{h}}_2^+\right]\right.$$
$$\cdot\left[\sum_{m=0}^{\infty}(ik_{2z})^m \int d\mathbf{k}'_\perp \mathbf{A}(\mathbf{k}'_\perp)Z'^{(m)}(\mathbf{k}_\perp - \mathbf{k}'_\perp)\right]$$
$$+ \left[\hat{\mathbf{h}}_2^+\hat{\mathbf{v}}_2^+ - \hat{\mathbf{v}}_2^+\hat{\mathbf{h}}_2^+\right] \cdot \left[\sum_{m=0}^{\infty}(ik_{2z})^m\right.$$
$$\left.\left.\int d\mathbf{k}'_\perp \mathbf{B}(\mathbf{k}'_\perp)Z'^{(m)}(\mathbf{k}_\perp - \mathbf{k}'_\perp)\right]\right\}, \quad \forall z > z_M, \quad (8.20)$$

Equations (8.19) and (8.20) must be solved to obtain the FTs of the surface currents; these can be substituted in Equations (8.14) and (8.15) to obtain the scattered and transmitted field, respectively. The main advantage of solving Equations (8.19) and (8.20) instead Equations (8.1) and (8.2) is that they do not involve the two space and spectral integrals; only one convolution is present between the FTs of unknown fields and the surface profile.

However, solution of Equations (8.19) and (8.20) is not straightforward at this stage. A perturbative approach is suggested by the SPM and is presented in the next Sections: within appropriate conditions, Equations (8.19) and (8.20) can be solved iteratively to obtain, to a generic order, the solutions for the FT of the surface currents: these are substituted in Equations (8.14) and (8.15) to obtain the corresponding order scattered and transmitted field, respectively.

8.4. Set up the Small Perturbation Method

A perturbation method can be applied to solve Equations (8.19) and (8.20) and obtained the scattered and transmitted fields via Equations (8.14) and (8.15). This method is based on the fulfilment of some conditions.

The first condition is set on the surface profile. First of all, the expansion in Equation (8.5) is truncated to a finite order M. To accept this approximation it is required that:

$$k_{(1,2)z}z' \ll 1, \quad (8.21)$$

which enforces a bounds on the surface heights variations in terms of the incident field wavelength. This first condition does not yet allow us to

8.4. Set up the Small Perturbation Method

solve Equations (8.19) and (8.20). In addition, it is necessary to introduce a second condition that also bounds the surface slopes:

$$\frac{\partial z'}{\partial x'} \ll 1, \frac{\partial z'}{\partial y'} \ll 1. \tag{8.22}$$

Use of both Conditions (8.21) and (8.22) is the prerequisite to develop an iterative solution to the scattering problem and justify the name, SPM, of the method whose rationale is illustrated hereafter.

Condition (8.21) suggests employing a coordinate system which includes $\hat{\mathbf{z}}$, the unit normal to the surface mean plane. The reason is the convenience to express the unknown surface fields $\mathbf{a}$ and $\mathbf{b}$, appearing in Equation (8.1) through (8.4), in terms of their transverse and z-components:

$$\begin{aligned} \mathbf{a}(\mathbf{r}'_\perp) &= \mathbf{a}_\perp \hat{\mathbf{r}}'_\perp + a_z \hat{\mathbf{z}} \\ \mathbf{b}(\mathbf{r}'_\perp) &= \mathbf{b}_\perp \hat{\mathbf{r}}'_\perp + b_z \hat{\mathbf{z}} \end{aligned} \tag{8.23}$$

Recalling that $\mathbf{a}$ and $\mathbf{b}$ are tangent to the surface, their z-component are expressed as:

$$\begin{aligned} a_z(\mathbf{r}'_\perp) &= \left(\hat{\mathbf{x}} \frac{\partial z'(\mathbf{r}'_\perp)}{\partial x} + \hat{\mathbf{y}} \frac{\partial z'(\mathbf{r}'_\perp)}{\partial y'} \right) \cdot \mathbf{a}_\perp(\mathbf{r}'_\perp) \\ b_z(\mathbf{r}'_\perp) &= \left(\hat{\mathbf{x}} \frac{\partial z'(\mathbf{r}'_\perp)}{\partial x'} + \hat{\mathbf{y}} \frac{\partial z'(\mathbf{r}'_\perp)}{\partial y'} \right) \cdot \mathbf{b}_\perp(\mathbf{r}'_\perp) \end{aligned} \tag{8.24}$$

see Equations (4.65) through (4.67).

The simple counterpart of Equations (8.23) and (8.24) in the Fourier domain is easy to obtain. The FT of Equations (8.23) leads to:

$$\begin{aligned} \mathbf{A}(\mathbf{k}_\perp) &= \mathbf{A}_\perp \hat{\mathbf{r}}'_\perp + A_z \hat{\mathbf{z}}, \\ \mathbf{B}(\mathbf{k}_\perp) &= \mathbf{B}_\perp \hat{\mathbf{r}}'_\perp + B_z \hat{\mathbf{z}}. \end{aligned} \tag{8.25}$$

whereas the FT of Equations (8.24) along with the Borel theorem provides:

$$\begin{aligned} A_z(\mathbf{k}_\perp) &= -i \int_{-\infty}^{\infty} d\mathbf{k}'_\perp (\mathbf{k}_\perp - \mathbf{k}'_\perp) \cdot \mathbf{A}_\perp(\mathbf{k}'_\perp) Z'(\mathbf{k}_\perp - \mathbf{k}'_\perp) \\ B_z(\mathbf{k}_\perp) &= -i \int_{-\infty}^{\infty} d\mathbf{k}'_\perp (\mathbf{k}_\perp - \mathbf{k}'_\perp) \cdot \mathbf{B}_\perp(\mathbf{k}'_\perp) Z'(\mathbf{k}_\perp - \mathbf{k}'_\perp) \end{aligned} \tag{8.26}$$

The series appearing in Equations (8.12) through (8.15) where each term is of higher order compared to the previous ones, see Conditions (8.21), suggests a similar series expansion of the unknowns **a** and **b**:

$$\mathbf{a}(\mathbf{r}'_\perp) = \sum_{m=0}^{\infty} \mathbf{a}^{(m)}(\mathbf{r}'_\perp)$$

$$\mathbf{b}(\mathbf{r}'_\perp) = \sum_{m=0}^{\infty} \mathbf{b}^{(m)}(\mathbf{r}'_\perp)$$

(8.27)

where $\mathbf{a}^{(m)}$ and $\mathbf{b}^{(m)}$ are the m-th order terms for the SPM solution for **a** and **b**, respectively. The reason is to perform all products, individuate terms of equal order and find the corresponding order solution.

Examination of Equation (8.24) shows that at any order, the z-components of **a** and **b** are of higher order compared to the transverse ones, because the small slope assumption for the scattering surface is enforced, see Condition (8.22). Accordingly, at the order $m = 0$ of the SPM approach, the z-components of **a** and **b** should be taken equal to zero,

$$a_z^{(0)}(\mathbf{r}'_\perp) = 0$$
$$b_z^{(0)}(\mathbf{r}'_\perp) = 0$$

(8.28)

so that the surface fields are coincident with those relevant to a flat surface.

By implementing the proposed recursive approach at any order $m \geq 1$, it turns out that Equation (8.24) leads to:

$$a_z^{(m)}(\mathbf{r}'_\perp) = \left(\hat{\mathbf{x}} \frac{\partial z'(\mathbf{r}'_\perp)}{\partial x'} + \hat{\mathbf{y}} \frac{\partial z'(\mathbf{r}'_\perp)}{\partial y'}\right) \cdot \mathbf{a}_\perp^{(m-1)}(\mathbf{r}'_\perp)$$

$$b_z^{(m)}(\mathbf{r}'_\perp) = \left(\hat{\mathbf{x}} \frac{\partial z'(\mathbf{r}'_\perp)}{\partial x'} + \hat{\mathbf{y}} \frac{\partial z'(\mathbf{r}'_\perp)}{\partial y'}\right) \cdot \mathbf{b}_\perp^{(m-1)}(\mathbf{r}'_\perp)$$

(8.29)

Similarly, in the recursive framework, Equations (8.26) takes the form:

$$A_z^{(m)}(\mathbf{k}_\perp) = -i \int_{-\infty}^{\infty} d\mathbf{k}'_\perp (\mathbf{k}_\perp - \mathbf{k}'_\perp) \cdot \mathbf{A}_\perp^{(m-1)}(\mathbf{k}'_\perp) Z'(\mathbf{k}_\perp - \mathbf{k}'_\perp)$$

$$B_z^{(m)}(\mathbf{k}_\perp) = -i \int_{-\infty}^{\infty} d\mathbf{k}'_\perp (\mathbf{k}_\perp - \mathbf{k}'_\perp) \cdot \mathbf{B}_\perp^{(m-1)}(\mathbf{k}'_\perp) Z'(\mathbf{k}_\perp - \mathbf{k}'_\perp)$$

(8.30)

These z-components appear as known terms in the solution of the equation, thus improving the method stability.

8.5. An Appropriate Coordinate System

In previous chapters the fields have been referred to as the coordinate system $(\hat{\mathbf{h}}, \hat{\mathbf{v}}, \hat{\mathbf{k}})$. Use of this reference system is less convenient here: the reason is that, in the SPM the transverse (to the z-axis) components of the fields are dominant compared to the longitudinal (along the z-axis) ones. Accordingly, a reference system that makes use of the unit vector $\hat{\mathbf{z}}$ is desired, so that equation processing and approximation implementation becomes more transparent. The new alternative coordinate system $(\hat{\mathbf{s}}, \hat{\mathbf{t}}, \hat{\mathbf{z}})$ is given by the projections of $\hat{\mathbf{v}}$ and $\hat{\mathbf{h}}$ orthogonal to $\hat{\mathbf{z}}$:

$$\hat{\mathbf{t}}(\mathbf{k}_\perp) = \hat{\mathbf{h}}^\pm_{(1,2)} \tag{8.31a}$$

$$\hat{\mathbf{s}}(\mathbf{k}_\perp) = \hat{\mathbf{t}}(\mathbf{k}_\perp) \times \hat{\mathbf{z}} = \hat{\mathbf{k}}_\perp, \tag{8.31b}$$

and the normal $\hat{\mathbf{z}}$ to the surface mean plane. This reference system is used to represent the unknown surface fields in the transformed domain:

$$\begin{aligned}\mathbf{A}(\mathbf{k}_\perp) &= A_t(\mathbf{k}_\perp)\hat{\mathbf{t}}(\mathbf{k}_\perp) + A_s(\mathbf{k}_\perp)\hat{\mathbf{s}}(\mathbf{k}_\perp) + A_z(\mathbf{k}_\perp)\hat{\mathbf{z}} \\ \mathbf{B}(\mathbf{k}_\perp) &= B_t(\mathbf{k}_\perp)\hat{\mathbf{t}}(\mathbf{k}_\perp) + B_s(\mathbf{k}_\perp)\hat{\mathbf{s}}(\mathbf{k}_\perp) + B_z(\mathbf{k}_\perp)\hat{\mathbf{z}}\end{aligned} \tag{8.32}$$

Moreover, the orthogonal components of the wavevectors can be written in terms of a polar coordinate system

$$\mathbf{k}_\perp = k_\rho \cos\phi_k \hat{\mathbf{x}} + k_\rho \sin\phi_k \hat{\mathbf{y}} \tag{8.33a}$$

$$\mathbf{k}'_\perp = k'_\rho \cos\phi'_k \hat{\mathbf{x}} + k'_\rho \sin\phi'_k \hat{\mathbf{y}}. \tag{8.33b}$$

The two coordinate systems $(\hat{\mathbf{s}}, \hat{\mathbf{t}}, \hat{\mathbf{z}})$ and $(\hat{\mathbf{h}}, \hat{\mathbf{v}}, \hat{\mathbf{k}})$ are linked. As a matter of fact:

$$\begin{aligned}\hat{\mathbf{v}}^\pm_{(1,2)} &= \hat{\mathbf{k}}^\pm_{(1,2)} \times \hat{\mathbf{h}}^\pm_{(1,2)} \\ &= \hat{\mathbf{k}}^\pm_{(1,2)} \times \hat{\mathbf{t}}(\mathbf{k}_\perp) = \mp \frac{k_{(1,2)z}}{k_{(1,2)}}\hat{\mathbf{s}}(\mathbf{k}_\perp) + \frac{k_\rho}{k_{(1,2)}}\hat{\mathbf{z}}.\end{aligned} \tag{8.34}$$

Equations (8.31a) and (8.34) are employed to evaluate the dot products in Equations (8.14), (8.15), (8.19) and (8.20) between the dyadics and the vectors representing the surface fields in the transformed domain:

$$\left(\hat{\mathbf{v}}^{\pm}_{(1,2)}\hat{\mathbf{v}}^{\pm}_{(1,2)} + \hat{\mathbf{h}}^{\pm}_{(1,2)}\hat{\mathbf{h}}^{\pm}_{(1,2)}\right) \cdot \mathbf{A}(\mathbf{k}_\perp)$$

$$= \hat{\mathbf{v}}^{\pm}_{(1,2)}\left[\frac{k_\rho}{k_{(1,2)}}A_z(\mathbf{k}_\perp) \mp \frac{k_{(1,2)z}}{k_{(1,2)}}A_s(\mathbf{k}_\perp)\right] + \hat{\mathbf{h}}^{\pm}_{(1,2)}A_t(\mathbf{k}_\perp) \quad (8.35)$$

$$\left(\hat{\mathbf{h}}^{\pm}_{(1,2)}\hat{\mathbf{v}}^{\pm}_{(1,2)} - \hat{\mathbf{v}}^{\pm}_{(1,2)}\hat{\mathbf{h}}^{\pm}_{(1,2)}\right) \cdot \mathbf{B}(\mathbf{k}_\perp)$$

$$= \hat{\mathbf{h}}^{\pm}_{(1,2)}\left[\frac{k_\rho}{k_{(1,2)}}B_z(\mathbf{k}_\perp) \mp \frac{k_{(1,2)z}}{k_{(1,2)}}B_s(\mathbf{k}_\perp)\right] - \hat{\mathbf{v}}^{\pm}_{(1,2)}B_t(\mathbf{k}_\perp). \quad (8.36)$$

8.6. Zero-order Solution

In order to get the solution up to the zero-order, the series appearing in Equations (8.6) through (8.9) are truncated at $M = 0$; and the similarly truncated FTs of Expansions (8.27) for the surface fields are substituted in the same equations. Then, only the zero-order terms must be retained. This implies that:

$$\begin{aligned} A_z^{(0)} &= 0 \\ B_z^{(0)} &= 0 \end{aligned}, \quad (8.37)$$

as it was already recognized, see Equation (8.28). The conclusion is that the zero-order surface fields in the transformed domain are orthogonal to the normal to the mean plane surface.

It is easy to check that, at this zero order, Equations (8.19) and (8.20) becomes the following ones:

$$\hat{\mathbf{e}}_i E_i \delta(\mathbf{k}_\perp - \mathbf{k}_{i\perp}) = \frac{1}{2}\frac{k_1}{k_{1z}}\left[\hat{\mathbf{v}}_1^-\hat{\mathbf{v}}_1^- + \hat{\mathbf{h}}_1^-\hat{\mathbf{h}}_1^-\right] \cdot \mathbf{A}_\perp^{(0)}(\mathbf{k}_\perp)$$

$$+ \frac{1}{2}\frac{k_1}{k_{1z}}\left[\hat{\mathbf{h}}_1^-\hat{\mathbf{v}}_1^- - \hat{\mathbf{v}}_1^-\hat{\mathbf{h}}_1^-\right] \cdot \mathbf{B}_\perp^{(0)}(\mathbf{k}_\perp) \quad (8.38)$$

$$0 = \frac{1}{2}\frac{k_1}{k_{2z}}\left[\hat{\mathbf{v}}_2^+\hat{\mathbf{v}}_2^+ + \hat{\mathbf{h}}_2^+\hat{\mathbf{h}}_2^+\right] \cdot \mathbf{A}_\perp^{(0)}(\mathbf{k}_\perp)$$

$$+ \frac{1}{2}\frac{k_2}{k_{2z}}\left[\hat{\mathbf{h}}_2^+\hat{\mathbf{v}}_2^+ - \hat{\mathbf{v}}_2^+\hat{\mathbf{h}}_2^+\right] \cdot \mathbf{B}_\perp^{(0)}(\mathbf{k}_\perp), \quad (8.39)$$

because Equation (8.11) provides $Z'^{(0)}(\mathbf{k}_\perp) = \delta(\mathbf{k}_\perp)$ and the integrals are easily evaluated. Solution of Equations (8.38) and (8.39) is initiated by

8.6. Zero-order Solution

enforcing the representation of the surface fields $\mathbf{A}_\perp^{(0)}(\mathbf{k}_\perp)$ and $\mathbf{B}_\perp^{(0)}(\mathbf{k}_\perp)$ in the reference system $(\hat{\mathbf{s}}, \hat{\mathbf{t}}, \hat{\mathbf{z}})$ orthogonal system. Furthermore, examination of Equation (8.38) suggests the following position:

$$\mathbf{A}_\perp^{(0)}(\mathbf{k}_\perp) = E_i \left(\hat{\mathbf{t}} a_t^{(0)} + \hat{\mathbf{s}} a_s^{(0)} \right) \delta(\mathbf{k}_\perp - \mathbf{k}_{i\perp})$$
$$\mathbf{B}_\perp^{(0)}(\mathbf{k}_\perp) = E_i \left(\hat{\mathbf{t}} b_t^{(0)} + \hat{\mathbf{s}} b_s^{(0)} \right) \delta(\mathbf{k}_\perp - \mathbf{k}_{i\perp})$$
(8.40)

that essentially implies selection of the wavevector characterised by

$$\mathbf{k}_\perp = \mathbf{k}_{i\perp};$$
(8.41)

in Equations (8.38) and (8.39), as well as in the zero-order version of Equations (8.14) and (8.15) that provide the zero-order scattered and transmitted field. Condition (8.41) also fixes the z-components of the wavevectors involved in the zero-order SPM solution:

$$k_{1z}^2 = k_{iz}^2 = k_1^2 - k_{i\perp}^2$$
$$k_{2z}^2 = k_2^2 - k_{i\perp}^2$$
(8.42)

Equations (8.40) are substituted into Equations (8.38) and (8.39), thus forming a linear system of two vector equations in two unknowns. To convert the vector equations into scalar ones, appropriate projections are required.

Equation (8.38) is projected onto $\hat{\mathbf{v}}_1^-$, and Equation (8.39) onto $\hat{\mathbf{v}}_2^+$, respectively, with the result:

$$\hat{\mathbf{v}}_1^- \cdot \hat{\mathbf{e}}_i = \frac{1}{2} a_s^{(0)} - \frac{1}{2} \frac{k_1}{k_{1z}} b_t^{(0)}$$
(8.43)

$$0 = -\frac{1}{2} \frac{k_1}{k_2} a_s^{(0)} - \frac{1}{2} \frac{k_2}{k_{2z}} b_t^{(0)}.$$
(8.44)

Equations (8.43) and (8.44) form a linear system with constant coefficients $a_s^{(0)}, b_t^{(0)}$ see Equations (8.41) and (8.42) to be solved in the unknown zero-order unknowns (a_s, b_t). The solution can be cast in the following form:

$$a_s^{(0)}(\mathbf{k}_\perp) = [\hat{\mathbf{v}}_1^- \cdot \hat{\mathbf{e}}_i](1 + R_{v0})$$
(8.45)

$$b_t^{(0)}(\mathbf{k}_\perp) = [\hat{\mathbf{v}}_1^- \cdot \hat{\mathbf{e}}_i] \frac{k_{1z}}{k_1}(1 - R_{v0})$$
(8.46)

wherein the factor

$$R_{v0} = \frac{k_2^2 k_{1z} - k_1^2 k_{2z}}{k_2^2 k_{1z} + k_1^2 k_{2z}}, \qquad (8.47)$$

has been introduced; by considering Equations (8.42) R_{v0} is recognized to be equal to the **v**-polarised Fresnel coefficient relevant to the surface mean plane.

Similarly, projections of Equation (8.38) onto $\hat{\mathbf{h}}_1^-$, and Equation (8.39) onto $\hat{\mathbf{h}}_2^+$ leads to:

$$\hat{\mathbf{h}}_1^- \cdot \hat{\mathbf{e}}_i = \frac{1}{2} \frac{k_1}{k_{1z}} a_t^{(0)} + \frac{1}{2} b_t^{(0)} \qquad (8.48)$$

$$0 = \frac{1}{2} \frac{k_1}{k_{2z}} a_t^{(0)} - \frac{1}{2} b_s^{(0)}. \qquad (8.49)$$

By considering Equations (8.41) and (8.42), the Equations (8.48) and (8.49) form a linear system with constant coefficients to be solved in the unknown zero-order unknowns $a_t^{(0)}, b_s^{(0)}$. The solution can be cast in the following form:

$$a_t^{(0)}(\mathbf{k}_\perp) = \left[\hat{\mathbf{h}}_1^- \cdot \hat{\mathbf{e}}_i\right] \frac{k_{1z}}{k_1} (1 - R_{h0}) \qquad (8.50)$$

$$b_s^{(0)}(\mathbf{k}_\perp) = \left[\hat{\mathbf{h}}_1^- \cdot \hat{\mathbf{e}}_i\right] (1 + R_{h0}), \qquad (8.51)$$

wherein the coefficient

$$R_{h0} = \frac{k_{1z} - k_{2z}}{k_{1z} + k_{2z}}, \qquad (8.52)$$

has been introduced; examination of Equations (8.42) R_{h0} is recognized to be equal to the **h**-polarised Fresnel coefficient relevant to the surface mean plane.

Substitution of the transformed domain values for the surface currents provided by Equations (8.40), (8.45) (8.46) (8.50) and (8.51) in the zero-order version of Equation (8.14)

$$\mathbf{E}^{(s,0)}(\mathbf{r}) = -\frac{1}{2} \int_{-\infty}^{\infty} d\mathbf{k}_\perp \exp(-i\mathbf{k}_\perp \cdot \mathbf{r}_\perp - ik_{1z}z)$$

$$\frac{k_1}{k_{1z}} \left\{ \left[\hat{\mathbf{v}}_1^+ \hat{\mathbf{v}}_1^+ + \hat{\mathbf{h}}_1^+ \hat{\mathbf{h}}_1^+\right] \cdot \mathbf{A}^{(0)}(\mathbf{k}'_\perp) \right.$$

$$\left. + \left[\hat{\mathbf{h}}_1^+ \hat{\mathbf{v}}_1^+ - \hat{\mathbf{v}}_1^+ \hat{\mathbf{h}}_1^+\right] \cdot \mathbf{B}^{(0)}(\mathbf{k}'_\perp) \right\}, \quad \forall z > z_M \qquad (8.53)$$

8.6. Zero-order Solution

leads to express, in closed form, the zero-order SPM solution for the scattered field:

$$\mathbf{E}^{(s,0)}(\mathbf{r}) = E_i \left[R_{h0} \hat{\mathbf{h}}_1^+ \left[\hat{\mathbf{h}}_1^- \cdot \hat{\mathbf{e}}_i \right] + R_{v0} \hat{\mathbf{v}}_1^+ \left[\hat{\mathbf{v}}_1^- \cdot \hat{\mathbf{e}}_i \right] \right]$$
$$\exp\left\{ -i \left[\mathbf{k}_{i\perp} \cdot \mathbf{r}_\perp + \left(k_1^2 - k_{i\perp}^2 \right) z \right] \right\}, \qquad (8.54)$$

wherein $\hat{\mathbf{h}}_1^-$ and $\hat{\mathbf{v}}_1^-$ take the values pertinent to the incident direction, whereas $\hat{\mathbf{h}}_1^+$ and $\hat{\mathbf{v}}_1^+$ take those for the specular direction.

Similarly, the zero-order version of Equation (8.15)

$$\mathbf{E}^{(t,0)}(\mathbf{r}) = \frac{1}{2} \int_{-\infty}^{\infty} d\mathbf{k}_\perp \exp(-i\mathbf{k}_\perp \cdot \mathbf{r}_\perp + ik_{2z}\hat{\mathbf{z}} \cdot \mathbf{r}) \frac{k_2}{k_{2z}} \left\{ \frac{k_1}{k_2} \left[\hat{\mathbf{v}}_2^- \hat{\mathbf{v}}_2^- + \hat{\mathbf{h}}_2^- \hat{\mathbf{h}}_2^- \right] \right.$$
$$\left. \cdot \mathbf{A}^{(0)}(\mathbf{k}'_\perp) + \left[\hat{\mathbf{h}}_2^- \hat{\mathbf{v}}_2^- - \hat{\mathbf{v}}_2^- \hat{\mathbf{h}}_2^- \right] \cdot \mathbf{B}^{(0)}(\mathbf{k}'_\perp) \right\}$$
$$(8.55)$$

is considered in order to evaluate the transmitted field. Proceeding along the same line, it turns out that:

$$\mathbf{E}^{(t,0)}(\mathbf{r}) = E_i \left[(1+R_h) \hat{\mathbf{h}}_2^- \left[\hat{\mathbf{h}}_1^- \cdot \hat{\mathbf{e}}_i \right] + (1+R_v) \frac{k_1}{k_2} \hat{\mathbf{v}}_2^+ \left[\hat{\mathbf{v}}_1^- \cdot \hat{\mathbf{e}}_i \right] \right]$$
$$\exp\left\{ -i \left[\mathbf{k}_{i\perp} \cdot \mathbf{r}_\perp + \left(k_2^2 - k_{i\perp}^2 \right) z \right] \right\}, \qquad (8.56)$$

wherein $\hat{\mathbf{h}}_1^-$ and $\hat{\mathbf{v}}_1^-$ take the values pertinent to the incident direction, whereas $\hat{\mathbf{h}}_2^+$ and $\hat{\mathbf{v}}_2^+$ take the values pertinent to the specular direction.

Equations (8.54) and (8.56) express the scattered and transmitted fields: note the presence of the Fresnel reflection coefficient R in the former and the transmission coefficient $1 + R$ in the latter expression. Also, factors appropriate to parallel and perpendicular polarisations appear at the right place. As for flat surfaces of infinite extent, a single plane wave is scattered in the specular direction and provides the coherent contribution to the scattered field. Similarly, a single plane wave is transmitted in the lower medium. For surfaces of finite extent, Equations (8.54) and (8.56) modify, and the zero-order field is scattered and transmitted in any direction with the usual sinc($\cdot$) shape. In both cases, no fractal parameter appears: accordingly, the solution is of deterministic nature being coincident with that of a plane surface.

8.7. First-order Solution

In order to get the solution up to the first order, the procedure introduced in Section 8.6 is systematically iterated. The terms of the first-order SPM solution are formally identified by the symbol 1 in the apex parenthesis. The series appearing in Equations (8.6) through (8.9) are truncated at $M = 1$ and the similarly truncated Expansions (8.27) are substituted therein. Then, only the terms of order 1 are individuated and retained in the considered equations. Accordingly, a system of four scalar equations is obtained in the four unknowns $\mathbf{A}_t^{(1)}(\mathbf{k}_\perp)$, $\mathbf{A}_s^{(1)}(\mathbf{k}_\perp)$, $\mathbf{B}_t^{(1)}(\mathbf{k}_\perp)$ and $\mathbf{B}_s^{(1)}(\mathbf{k}_\perp)$. The procedure can be systematically implemented, finally leading to the evaluation of the transverse (to the z-axis) components of the field.

Before proceeding further, it is noted that, differently from the zero-order SPM solution, the first-order one exhibits a z-component for the surface fields. By benefiting of Condition (8.22), these components can be evaluated by setting $m = 1$ in Equations (8.30), and by substituting the zero-order solution for the transverse surface fields that has been obtained in the previous step and whose formal expression is reported in Equations (8.40). It turns out that:

$$A_z^{(1)}(\mathbf{k}_\perp) = -iE_iZ'(\mathbf{k}_\perp - \mathbf{k}_{i\perp})(\mathbf{k}_\perp - \mathbf{k}_{i\perp}) \cdot (\hat{\mathbf{t}} a_t^{(0)} + \hat{\mathbf{s}} a_s^{(0)})$$
$$B_z^{(1)}(\mathbf{k}_\perp) = -iE_iZ'(\mathbf{k}_\perp - \mathbf{k}_{i\perp})(\mathbf{k}_\perp - \mathbf{k}_{i\perp}) \cdot (\hat{\mathbf{t}} b_t^{(0)} + \hat{\mathbf{s}} b_s^{(0)}) \quad , \quad (8.57)$$

wherein Definition (8.11) leads to set $Z'^{(1)}(\mathbf{k}_\perp - \mathbf{k}'_\perp) = Z'(\mathbf{k}_\perp - \mathbf{k}'_\perp)$.

Solution of Equations (8.19) and (8.20) aimed to provide the FT of the surface currents is now in order. Their first order component is obtained by following the rationale presented in Section 8.2 and the set up reported in Section 8.4. Accordingly, the first order version of Equations (8.19) and (8.20) is given by:

$$\frac{1}{2}\frac{k_1}{k_{1z}}\left\{\left[\hat{\mathbf{v}}_1^-\hat{\mathbf{v}}_1^- + \hat{\mathbf{h}}_1^-\hat{\mathbf{h}}_1^-\right] \cdot \mathbf{A}_\perp^{(1)}(\mathbf{k}_\perp) + \left[\hat{\mathbf{h}}_1^-\hat{\mathbf{v}}_1^- - \hat{\mathbf{v}}_1^-\hat{\mathbf{h}}_1^-\right] \cdot \mathbf{B}_\perp^{(1)}(\mathbf{k}_\perp)\right\}$$
$$= \frac{1}{2}\frac{k_1}{k_{1z}}iE_iZ'(\mathbf{k}_\perp - \mathbf{k}_{i\perp})\left\{\left[\hat{\mathbf{v}}_1^-\hat{\mathbf{v}}_1^- + \hat{\mathbf{h}}_1^-\hat{\mathbf{h}}_1^-\right] \cdot (\mathbf{k}_\perp - \mathbf{k}_{i\perp})\right.$$
$$\left.\cdot (\hat{\mathbf{t}} a_t^{(0)} + \hat{\mathbf{s}} a_s^{(0)})\hat{\mathbf{z}} + k_{1z}(\hat{\mathbf{t}} a_t^{(0)} + \hat{\mathbf{s}} a_s^{(0)}) + \left[\hat{\mathbf{h}}_1^-\hat{\mathbf{v}}_1^- - \hat{\mathbf{v}}_1^-\hat{\mathbf{h}}_1^-\right]\right.$$
$$\left.\cdot (\mathbf{k}_\perp - \mathbf{k}_{i\perp}) \cdot (\hat{\mathbf{t}} b_t^{(0)} + \hat{\mathbf{s}} b_s^{(0)})\hat{\mathbf{z}} + k_{1z}(\hat{\mathbf{t}} b_t^{(0)} + \hat{\mathbf{s}} b_s^{(0)})\right\}, \quad (8.58)$$

$$\frac{1}{2}\frac{k_2}{k_{2z}}\left\{\frac{k_1}{k_2}\left[\hat{\mathbf{v}}_2^+\hat{\mathbf{v}}_2^+ + \hat{\mathbf{h}}_2^+\hat{\mathbf{h}}_2^+\right]\cdot\mathbf{A}_\perp^{(1)}(\mathbf{k}_\perp) + \left[\hat{\mathbf{h}}_2^+\hat{\mathbf{v}}_2^+ - \hat{\mathbf{v}}_2^+\hat{\mathbf{h}}_2^+\right]\cdot\mathbf{B}_\perp^{(1)}(\mathbf{k}_\perp)\right\}$$

$$= \frac{1}{2}\frac{k_2}{k_{2z}}iE_iZ'(\mathbf{k}_\perp - \mathbf{k}_{i\perp})\left\{\frac{k_1}{k_2}\left[\hat{\mathbf{v}}_2^+\hat{\mathbf{v}}_2^+ + \hat{\mathbf{h}}_2^+\hat{\mathbf{h}}_2^+\right]\cdot(\mathbf{k}_\perp - \mathbf{k}_{i\perp})\right.$$

$$\cdot(\hat{\mathbf{t}}a_t^{(0)} + \hat{\mathbf{s}}a_s^{(0)})\hat{\mathbf{z}} - k_{2z}(\hat{\mathbf{t}}a_t^{(0)} + \hat{\mathbf{s}}a_s^{(0)}) + \left[\hat{\mathbf{h}}_2^+\hat{\mathbf{v}}_2^+ - \hat{\mathbf{v}}_2^+\hat{\mathbf{h}}_2^+\right]$$

$$\left.\cdot(\mathbf{k}_\perp - \mathbf{k}_{i\perp})\cdot(\hat{\mathbf{t}}b_t^{(0)} + \hat{\mathbf{s}}b_s^{(0)})\hat{\mathbf{z}} - k_{2z}(\hat{\mathbf{t}}b_t^{(0)} + \hat{\mathbf{s}}b_s^{(0)})\right\}.$$

(8.59)

To get Equations (8.58) and (8.59) two main step have been implemented: first, Equations (8.57) have been employed to express the first order z-component for the surface fields in terms of the zero-order ones; then, the convolution integral between the FT of the surface profile and the zero-order surface fields has been solved by benefiting of the solutions provided in Equations (8.40). This second step is easily accomplished due to the simple integral relation provided by the convolution form that has been obtained by employing Condition (8.21).

Equations (8.58) and (8.59) must be solved for the first order unknown orthogonal components $\mathbf{A}_\perp^{(1)}(\mathbf{k}_\perp)$ and $\mathbf{B}_\perp^{(1)}(\mathbf{k}_\perp)$ of the transformed surface fields. However, a simple inspection of Equations (8.58) and (8.59) leads to derive a number of interesting results. Equations (8.58) and (8.59) show that $\mathbf{A}_\perp^{(1)}(\mathbf{k}_\perp)$ and $\mathbf{B}_\perp^{(1)}(\mathbf{k}_\perp)$ are both proportional to the FT of the surface profile $Z'(\mathbf{k}_\perp - \mathbf{k}_{i\perp})$. More specifically, the FT of the surface fields at the wavevector individuated by the value of $\mathbf{k}_\perp$ is proportional to the amplitude of the FT of the surface profile at (the resonant) wavevector $(\mathbf{k}_\perp - \mathbf{k}_{i\perp})$.

Similarly to what done for the zero order solution, the vector Equations (8.58) and (8.59) are projected onto the $(\hat{\mathbf{s}}, \hat{\mathbf{t}}, \hat{\mathbf{z}})$ coordinate system, thus providing a system of four scalar equations to be solved in the four unknowns $A_t^{(1)}(\mathbf{k}_\perp)$, $A_s^{(1)}(\mathbf{k}_\perp)$, $B_t^{(1)}(\mathbf{k}_\perp)$ and $B_s^{(1)}(\mathbf{k}_\perp)$. The remaining z-components of the transformed surface fields, $A_z^{(1)}(\mathbf{k}_\perp)$ and $B_z^{(1)}(\mathbf{k}_\perp)$, are computed via Equations (8.57). The obtained transverse components, $\mathbf{A}_\perp^{(1)}(\mathbf{k}_\perp)$ and $\mathbf{B}_\perp^{(1)}(\mathbf{k}_\perp)$, of the FT of the surface fields, together with their associate z-components, are substituted in the first-order SPM equation for the scattered and transmitted field, thus formally leading to:

$$\mathbf{E}^{(s,1)}(\mathbf{r}) = -\frac{1}{2} \int_{-\infty}^{\infty} d\mathbf{k}_\perp \exp(-i\mathbf{k}_\perp \cdot \mathbf{r}_\perp - ik_{1z}z) \frac{k_1}{k_{1z}} \left\{ \left[\hat{\mathbf{v}}_1^+ \hat{\mathbf{v}}_1^+ + \hat{\mathbf{h}}_1^+ \hat{\mathbf{h}}_1^+ \right] \right.$$
$$\cdot \left[\mathbf{A}^{(1)}(\mathbf{k}_\perp) - ik_{1z} \int d\mathbf{k}'_\perp \mathbf{A}_\perp^{(0)}(\mathbf{k}'_\perp) Z'\left(\mathbf{k}_\perp - \mathbf{k}'_\perp\right) \right]$$
$$+ \left[\hat{\mathbf{h}}_1^+ \hat{\mathbf{v}}_1^+ - \hat{\mathbf{v}}_1^+ \hat{\mathbf{h}}_1^+ \right] \cdot \left[\mathbf{B}^{(1)}(\mathbf{k}_\perp) - ik_{1z} \right.$$
$$\left. \left. \int d\mathbf{k}'_\perp \mathbf{B}_\perp^{(0)}(\mathbf{k}'_\perp) Z'(\mathbf{k}_\perp - \mathbf{k}'_\perp) \right] \right\},$$

(8.60)

$$\mathbf{E}^{(t,1)}(\mathbf{r}) = \frac{1}{2} \int_{-\infty}^{\infty} d\mathbf{k}_\perp \exp(-i\mathbf{k}_\perp \cdot \mathbf{r}_\perp + ik_{2z}\hat{\mathbf{z}} \cdot \mathbf{r}) \frac{k_2}{k_{2z}} \left\{ \frac{k_1}{k_2} \left[\hat{\mathbf{v}}_2^+ \hat{\mathbf{v}}_2^- + \hat{\mathbf{h}}_2^- \hat{\mathbf{h}}_2^- \right] \right.$$
$$\cdot \left[\mathbf{A}^{(1)}(\mathbf{k}_\perp) + ik_{2z} \int d\mathbf{k}'_\perp \mathbf{A}_\perp^{(0)}(\mathbf{k}'_\perp) Z'(\mathbf{k}_\perp - \mathbf{k}'_\perp) \right]$$
$$+ \left[\hat{\mathbf{h}}_2^- \hat{\mathbf{v}}_2^- - \hat{\mathbf{v}}_2^- \hat{\mathbf{h}}_2^- \right] \cdot \left[\mathbf{B}^{(1)}(\mathbf{k}_\perp) + ik_{2z} \right.$$
$$\left. \left. \int d\mathbf{k}'_\perp \mathbf{B}_\perp^{(0)}(\mathbf{k}'_\perp) Z'(\mathbf{k}_\perp - \mathbf{k}'_\perp) \right] \right\}.$$

(8.61)

Equations (8.40) and (8.57) are then substituted in Equations (8.60) and (8.61) to provide a convenient first-order SPM expression for the scattered and transmitted fields:

$$\mathbf{E}^{(s,1)}(\mathbf{r}) = -\frac{1}{2} \int_{-\infty}^{\infty} d\mathbf{k}_\perp \exp(-i\mathbf{k}_\perp \cdot \mathbf{r}_\perp - ik_{1z}z) \frac{k_1}{k_{1z}} \left\{ \left[\hat{\mathbf{v}}_1^+ \hat{\mathbf{v}}_1^+ + \hat{\mathbf{h}}_1^+ \hat{\mathbf{h}}_1^+ \right] \right.$$
$$\cdot \left[\mathbf{A}_\perp^{(1)}(\mathbf{k}_\perp) - iE_i Z'(\mathbf{k}_\perp - \mathbf{k}_{i\perp}) \left[(\mathbf{k}_\perp - \mathbf{k}_{i\perp}) \cdot (\hat{\mathbf{t}} a_t^{(0)} + \hat{\mathbf{s}} a_s^{(0)}) \hat{\mathbf{z}} \right. \right.$$
$$\left. \left. + k_{1z} (\hat{\mathbf{t}} a_t^{(0)} + \hat{\mathbf{s}} a_s^{(0)}) \right] \right] + \left[\hat{\mathbf{h}}_1^+ \hat{\mathbf{v}}_1^+ - \hat{\mathbf{v}}_1^+ \hat{\mathbf{h}}_1^+ \right]$$
$$\cdot \left[\mathbf{B}_\perp^{(1)}(\mathbf{k}_\perp) - iE_i Z'(\mathbf{k}_\perp - \mathbf{k}_{i\perp}) \left[(\mathbf{k}_\perp - \mathbf{k}_{i\perp}) \right. \right.$$
$$\left. \left. \left. \cdot (\hat{\mathbf{t}} b_t^{(0)} + \hat{\mathbf{s}} b_s^{(0)}) \hat{\mathbf{z}} + k_{1z} (\hat{\mathbf{t}} b_t^{(0)} + \hat{\mathbf{s}} b_s^{(0)}) \right] \right] \right\},$$

(8.62)

$$\mathbf{E}^{(t,1)}(\mathbf{r}) = \frac{1}{2}\int_{-\infty}^{\infty} d\mathbf{k}_\perp \exp\left(-i\mathbf{k}_\perp \cdot \mathbf{r}_\perp + ik_{2z}\hat{\mathbf{z}}\cdot\mathbf{r}\right)\frac{k_2}{k_{2z}}\left\{\frac{k_1}{k_2}\left[\hat{\mathbf{v}}_2^-\hat{\mathbf{v}}_2^- + \hat{\mathbf{h}}_2^-\hat{\mathbf{h}}_2^-\right]\right.$$

$$\cdot\left[\mathbf{A}_\perp^{(1)}(\mathbf{k}_\perp) - iE_i Z'(\mathbf{k}_\perp - \mathbf{k}_{i\perp})\left[(\mathbf{k}_\perp - \mathbf{k}_{i\perp})\cdot(\hat{\mathbf{t}}a_t^{(0)} + \hat{\mathbf{s}}a_s^{(0)})\hat{\mathbf{z}}\right.\right.$$

$$\left.\left.-k_{2z}(\hat{\mathbf{t}}a_t^{(0)} + \hat{\mathbf{s}}a_s^{(0)})\right]\right] + \left[\hat{\mathbf{h}}_2^-\hat{\mathbf{v}}_2^- - \hat{\mathbf{v}}_2^-\hat{\mathbf{h}}_2^-\right]$$

$$\cdot\left[\mathbf{B}_\perp^{(1)}(\mathbf{k}_\perp) - iE_i Z'(\mathbf{k}_\perp - \mathbf{k}_{i\perp})[(\mathbf{k}_\perp - \mathbf{k}_{i\perp})\right.$$

$$\left.\left.\cdot(\hat{\mathbf{t}}b_t^{(0)} + \hat{\mathbf{s}}b_s^{(0)})\hat{\mathbf{z}} - k_{2z}(\hat{\mathbf{t}}b_t^{(0)} + \hat{\mathbf{s}}b_s^{(0)})\right]\right\}. \quad (8.63)$$

Equations (8.62) and (8.63) formally express, in a meaningful way, the scattered and transmitted fields in terms of the FT of the surface. It has already been noted that the unknown fields are proportional to $Z'(\mathbf{k}_\perp - \mathbf{k}_{i\perp})$. Examination of Equations (8.62) and (8.63) shows that this happens also for any other of the factors in the integrals. Asymptotic evaluation of the integrals at large distance shows that the scattered field along the generic direction individuated by $\mathbf{k}_{s\perp}$ is proportional to the integrand evaluated for $\mathbf{k}_\perp = \mathbf{k}_{s\perp}$. This implies that only the line $\mathbf{k}_{s\perp} - \mathbf{k}_{i\perp}$ of the spectrum $Z'(\mathbf{k}_\perp)$ of $z'(\mathbf{r}'_\perp)$ contributes to the field.

Computation of the scattered power-density in the far field, up to the first order, requires that the sum of the zero- and first-order terms of the field is considered. However, it has already been noted that the zero-order term is a deterministic one: accordingly, the statistical mean of the mixed product is proportional to the statistical average of $Z'(\mathbf{k}_\perp - \mathbf{k}_{i\perp})$, which approaches zero as the roughness is at least comparable to the wavelength. Accordingly, the statistical part of the scattered power density in the far field coincides with its first-order contribution.

To obtain this scattered power density, Equation (8.62) must be multiplied by its complex conjugate and the ensemble average must be implemented. These operations show that, in the direction individuated by the value of $\mathbf{k}_\perp = \mathbf{k}_{s\perp}$, the scattered power density is proportional to the amplitude of the surface power density spectrum at (the resonant) wavector $\mathbf{k}_{s\perp} - \mathbf{k}_{i\perp}$: this result is in agreement with the Bragg theory.

In this averaging procedure, it is convenient to recast the equation for the scattered power density in terms of the incident and the scattered polarisations p and q and the parameters for the scattering geometry depicted in Figure 8.1. The final result is to express the scattered power-density in terms of four scalar coefficients. The SPM expression of the scattered power-density turns out to be written as:

$$\left\langle \left|E_{pq}^{(s,1)}\right|^2\right\rangle = \frac{4A|E_i|^2 k^4 \cos^2\vartheta_s \cos^2\vartheta_i |G_{pq}|^2 W(\eta_{xy})}{(2\pi r)^2}, \quad (8.64)$$

that, for an isotropic surface can be expressed in terms of the fractal parameters introduced in Chapter 3 as:

$$\left\langle \left|E_{pq}^{(s,1)}\right|^2\right\rangle$$
$$= \frac{2^{2H+1}\Gamma^2(1+H)\sin(\pi H)A\cos^2\vartheta_s \cos^2\vartheta_i |E_i|^2 |G_{pq}|^2 (kT)^{2(1-H)}}{(\pi kr)^2(\sqrt{\sin^2\vartheta_i + \sin^2\vartheta_s - 2\sin\vartheta_i \sin\vartheta_s \cos\varphi_s})^{2H+2}}. \quad (8.65)$$

In Equations (8.53) and (8.54) G_{pq} is a coefficient that depends on polarisation, incidence and scattering angles, and the relative complex dielectric constant, ε_r, of the scattering medium:

$$G_{hh} = \frac{(\varepsilon_r - 1)\cos\varphi_s}{\left(\cos\vartheta_s + \sqrt{\varepsilon_r - \sin^2\vartheta_s}\right)\left(\cos\vartheta_i + \sqrt{\varepsilon_r - \sin^2\vartheta_i}\right)} \quad (8.66)$$

$$G_{hv} = \frac{(\varepsilon_r - 1)\sqrt{\varepsilon_r - \sin^2\vartheta_s}\sin\varphi_s}{\left(\cos\vartheta_s + \sqrt{\varepsilon_r - \sin^2\vartheta_s}\right)\left(\varepsilon_r \cos\vartheta_i + \sqrt{\varepsilon_r - \sin^2\vartheta_i}\right)} \quad (8.67)$$

$$G_{vh} = \frac{(\varepsilon_r - 1)\sqrt{\varepsilon_r - \sin^2\vartheta_s}\sin\varphi_s}{\left(\varepsilon_r \cos\vartheta_s + \sqrt{\varepsilon_r - \sin^2\vartheta_s}\right)\left(\cos\vartheta_i + \sqrt{\varepsilon_r - \sin^2\vartheta_i}\right)} \quad (8.68)$$

$$G_{vv} = \frac{\left(\sqrt{\varepsilon_r - \sin^2\vartheta_s}\sqrt{\varepsilon_r - \sin^2\vartheta_i}\cos\varphi_s - \varepsilon_r \sin\vartheta_i \sin\vartheta_s\right)(\varepsilon_r - 1)}{\left(\varepsilon_r \cos\vartheta_s + \sqrt{\varepsilon_r - \sin^2\vartheta_s}\right)\left(\varepsilon_r \cos\vartheta_i + \sqrt{\varepsilon_r - \sin^2\vartheta_i}\right)}. \quad (8.69)$$

8.8. Small Perturbation Method Limits of Validity

A priori conditions to apply the SPM have been introduced in Equations (8.21) and (8.22) for a deterministic surface. In the case of bandlimited

8.8. Small Perturbation Method Limits of Validity

fBm stochastic process, Eq. (8.21) takes the form:

$$k_{(1,2)z}\sigma \ll 1. \tag{8.70}$$

Similarly, Equation (8.22) takes the form:

$$\sigma' \ll 1. \tag{8.71}$$

Equations (8.74) and (8.75) can be expressed in terms of the fractal parameters by taking into account the results reported in Table 3.3. Thus, Conditions (8.74) and (8.75) read as

$$k_{(1,2)z}2^{2H-1}\Gamma^2(1+H)\frac{\sin(\pi H)}{\pi H}T^2\left[\left(\frac{\tau_M}{T}\right)^{2H} - \left(\frac{\tau_m}{T}\right)^{2H}\right] \ll 1, \tag{8.72}$$

$$2^{2H-1}\Gamma^2(1+H)\frac{\sin(\pi H)}{\pi(1-H)}\left[\left(\frac{T}{\tau_m}\right)^{2-2H} - \left(\frac{T}{\tau_M}\right)^{2-2H}\right] \ll 1, \tag{8.73}$$

respectively. The expression of the two parameters, τ_m, τ_M, appearing in Equations (8.72) and (8.73) are provided by Equations (3.48) and (3.49) as function of the surface dimensions and the incident wavelength.

For surfaces of infinite extent the zero-order solution provides a scattered field in the specular direction only; for surface of finite extent the zero-order solution for the scattered fields leads to the coherent component which takes the usual sinc(·) shape.

As far as the first order SPM solution is taken into account, for the mathematical fBm, the fBm power spectrum has been substituted in Equation (8.64) to get Equation (8.65). However, the considerations reported in Chapter 3 show that the power-law form for the power spectrum of the physical fBm can be fully applied only in an appropriate wavenumber interval $\kappa = \eta_{xy} \in [\kappa_m, \kappa_M]$; by considering Equations (3.48) and (3.49), this interval can be written in terms of the surface and viewing parameters:

$$k\sqrt{\sin^2\vartheta_i + \sin^2\vartheta_s - 2\sin\vartheta_i\sin\vartheta_s\cos\varphi_s} \in \left[\frac{2\pi\chi_1}{\sqrt{X^2+Y^2}}, \frac{2\pi}{\chi_2\lambda}\right]. \tag{8.74}$$

Then, Equation (8.74) provides a further validity limit, in terms of the scattering angles, to employ the first order SPM solution. In particular, Condition (8.74) shows that for surfaces of finite extent Equation (8.65) cannot be employed in the specular direction which is characterised by $\mathbf{k}_{s\perp} = \mathbf{k}_{i\perp}$: in this case the zero-order solution (8.54) can be used, providing that, as noted at the end of Section 8.6, the usual sinc(·) shape is assumed.

8.9. Influence of Fractal and Electromagnetic Parameters Over the Scattered Field

In this section the influence of fractal parameters, S_0 and H, on the scattered field is analyzed; for the sake of completeness the influence of the electromagnetic wavelength is also discussed. The scattering surface is illuminated from the direction $\vartheta_i = \frac{\pi}{4}$, whereas the scattered field is displayed for any ϑ_s in the plane identified by $\varphi_s = 0$; this suggests assuming $-\frac{\pi}{2} \leq \vartheta_s \leq \frac{\pi}{2}$. Scattered power density diagrams for the zero-order SPM solution are not displayed: their contents are trivial, being coincident with those relevant to a flat surface.

Scattered power density diagrams are displayed for the first order SPM solution only: the shape of these diagrams depends on the surface fractal parameters, but suffer for the infrared catastrophe that leads to a power density spectrum for the surface profile that diverges in the origin. Then, as shown by Condition (8.74), for scattering directions close to the specular one, the scattered power-density cannot be obtained by the first-order SPM solution, whereas the zero-order one can be conveniently employed. Moreover, by considering surface of not marginal roughness, for scattering directions not close to the specular one, the zero-order SPM solution becomes negligible when compared to the first-order one.

The effect of the fractal parameters on the scattered field is first briefly described in a qualitative manner. Then, quantitative assessment via Equations (8.65) is graphically presented in Figures 8.2 through 8.5. The presence of the factor $G_{pq}(\cdot)$, which depends on the electromagnetic parameters and fields polarisations is omitted: the reason is that it also appears if classical surfaces are considered, it does not include any dependence on the fractal parameters, and is not relevant to this discussion. Moreover, the scattered power density in each figure provides the scattering diagram only for angles not very close to the specular direction, see Condition (8.74).

The reference scattering surface is characterised by the fractal parameters $S_0 = 0.003076$ m^{2-2H} and $H = 0.8$. These parameters are typical for natural surfaces on the Earth when illuminated by a microwave instrument. Moreover, with the chosen parameters, this reference fBm process holds the same fractal parameters that are held by the WM reference surface (with $\nu = e$) presented in Chapter 3 and used in Chapter 5. Illumination and reference surface parameters are reported in Table 8.1. The corresponding scattered normalized power density is displayed in Figure 8.2. Then, in

8.9. Influence of Fractal and Electromagnetic Parameters

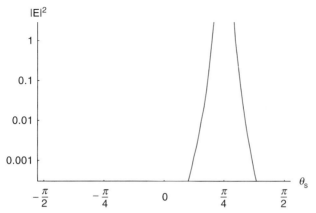

FIGURE 8.2 Plots of the normalised power density scattered by the reference surface holding the fractal parameters $H = 0.8$ and $S_0 = 0.003076$ m^{2-2H}. Illumination conditions and surface parameters for the reference surface are reported in Table 8.1.

TABLE 8.1 Parameters relevant to the illumination conditions and to the reference surface considered in Figure 8.2 and used to compare results in Figures 8.2 through 8.5. These parameters provide a reference surface corresponding to that employed in Chapter 5 where the WM model is used.

Incidence angle (ϑ_i)	45°
Illuminated area (X,Y)	(1m, 1m)
Electromagnetic wavelength (λ)	0.1 m
Hurst exponent (H)	0.8
Spectral amplitude (S_0)	0.003076 m^{2-2H}

any subsequent subsection only one of the surface fractal parameters is changed within a reasonable and significant range; its specific contribution to the scattered field is graphically displayed. The plot in Figure 8.2 shows that the maximum field scattered by the reference surface is attained in the specular direction. It is also evident that the average value evaluated for the scattered power density leads to a very smooth graph.

8.9.1. The Role of the Spectral Amplitude

The spectral amplitude S_0, directly influences the surface roughness, In Figure 8.3 two graphs are reported relevant to the power density scattered by surfaces characterized by different values of the spectral amplitude

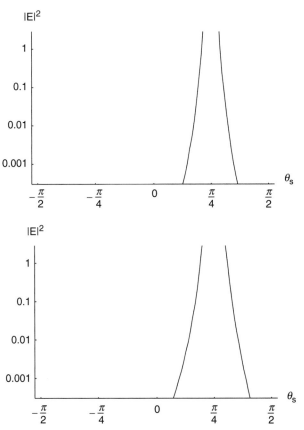

FIGURE 8.3 Modified versions of the reference surface are considered. As in Figure 8.2, corresponding normalized scattered power densities are depicted. Top $S_0 = 0.003076/2$ m^{2-2H}; bottom $S_0 = 0.003076 \cdot 2$ m^{2-2H}.

$S_0 = 0.003076/2$ m^{2-2H} and $S_0 = 0.003076 \cdot 2$ m^{2-2H}, one-half and double the value selected for the reference surface, respectively.

Comparison of Figures 8.2 and 8.3 shows that the smaller the spectral amplitude S_0, the narrower the power density diagram around the specular direction, and the higher the power density scattered in the specular direction.

8.9.2. The Role of the Hurst Exponent

As shown in Chapter 3, the Hurst coefficient is related to the fractal dimension. The lower H, the higher the fractal dimension.

8.9. Influence of Fractal and Electromagnetic Parameters

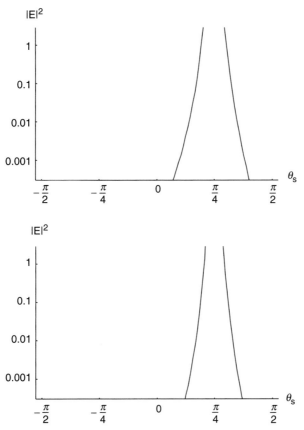

FIGURE 8.4 Modified versions of the reference surface are considered. As in Figure 8.2, corresponding normalized scattered power densities are depicted. Top $H = 0.7$; bottom $H = 0.9$.

In Figure 8.4 two graphs are reported, relevant to surfaces with higher, $H = 0.9$, and lower, $H = 0.7$, values of the Hurst coefficient with respect to the intermediate one $H = 0.8$ reported in Figure 8.2 and relevant to the reference surface. It is here recalled that changing the value of H, while leaving the numerical value of S_0 constant, corresponds to considering constant the amplitude of power density spectrum at unitary wavenumber.

Comparison of Figures 8.2 and 8.4 shows that the smaller the Hurst coefficient the narrower the power density diagram around the specular direction, and the higher the power density scattered in the specular direction.

8.9.3. The Role of the Electromagnetic Wavelength

The scattered field evaluated by means of PO solution and relevant to different values of λ is reported under Figure 8.5

In Figure 8.5 two graphs are reported, relevant to electromagnetic wavelengths with lower, $\lambda = 0.05$ m, and higher, $\lambda = 0.2$ m, values with respect to the intermediate one $\lambda = 0.1$ m reported in Figure 8.2 and relevant to the reference surface. Visual inspection of Equation (8.65) shows that the dependence of the scattered power density on the

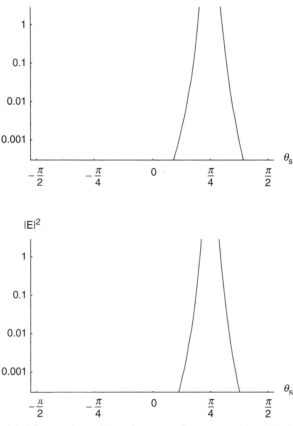

FIGURE 8.5 Modified versions of the reference surface are considered. As in Figure 8.2, corresponding normalized scattered power densities are depicted. Top $\lambda = 0.05$ m; bottom $\lambda = 0.2$ m.

electromagnetic wavelength can also be employed to study the dependence on the Topothesy, T.

Comparison of Figures 8.2 and 8.5 shows that the smaller the electromagnetic wavelength the broader the power density diagram around the specular direction, and the lower the power density scattered in the specular direction.

8.10. References and Further Readings

As far as classical surfaces are concerned the electromagnetic scattering evaluated by means of the SPM approach is conveniently reported in the books of Tsang (1985), Tsang (2001), Ulaby (1982). The second order SPM solution for classical surfaces is presented in the book of Tsang (2001). Connection between the SPM solution and the Integral Equation Approach is provided in the book by Fung (1994). A chapter in the book of Falconer (1990) is devoted to fBm surfaces along with relevant definitions and properties. The list of the papers written by the authors of this book and concerning the subject of this chapter is found in Appendix C.

Appendix A
Mathematical Formulae

Mathematical formulae are reported in this section as long as they are used in the corresponding chapters.

Chapter 2

$$\int_0^{2\pi} \exp[iz\cos\varphi]\exp[iv\varphi]\,d\varphi = (-i)^v\,2\pi J_v(z) \qquad (A.2.1)$$

$$\int_{-\infty}^{\infty} \frac{1}{\sqrt{2\pi}\sigma}\exp\left(-\frac{z^2}{2\sigma^2}\right)\exp(-i\xi z)\,dz = \exp\left(-\frac{1}{2}\sigma^2\xi^2\right) \qquad (A.2.2)$$

Chapter 3

$$\int_0^\infty x^\mu J_v(\tau x)\,dx = 2^\mu \tau^{-\mu-1}\frac{\Gamma\left(\tfrac{1}{2}+\tfrac{1}{2}v+\tfrac{1}{2}\mu\right)}{\Gamma\left(\tfrac{1}{2}+\tfrac{1}{2}v-\tfrac{1}{2}\mu\right)},$$
$$(-v-1 < \mu < 0.5 \text{ and } \tau > 0) \qquad (A.3.1)$$

$$z\Gamma(z) = \Gamma(1+z) \qquad (A.3.2)$$

$$\Gamma(1+z)\Gamma(1-z) = \frac{\pi z}{\sin(\pi z)} \qquad (A.3.3)$$

Chapter 4

$$\nabla \cdot [(\nabla \times \mathbf{A}) \times \mathbf{B} + \mathbf{A} \times (\nabla \times \mathbf{B})] = \nabla \times \nabla \times \mathbf{A} \cdot \mathbf{B} - \mathbf{A} \cdot \nabla \times \nabla \times \mathbf{B}$$
(A.4.1)

Chapter 5

$$\exp\{ia \sin \xi\} = \sum_{m=-\infty}^{+\infty} J_m(a) \exp\{im\xi\}$$
(A.5.1)

$$\prod_{p=0}^{P-1} \sum_{m=-\infty}^{+\infty} J_m(a_p) \exp\{im\xi_p\}$$

$$= \sum_{m_0=-\infty}^{+\infty} \cdots \sum_{m_{P-1}=-\infty}^{+\infty} \left[\exp\left(i \sum_{p=0}^{P-1} m_p \xi_p\right) \prod_{p=0}^{P-1} J_{m_p}(a_p) \right]$$
(A.5.2)

Chapter 6

$$J_0(t) = \sum_{n=0}^{\infty} (-1)^n \frac{t^{2n}}{2^{2n}(n!)^2}$$
(A.6.1)

$$\int_0^\infty \exp(-u\tau^v) \cdot \tau^w d\tau = \frac{1}{v} \cdot \frac{1}{u^{\frac{w+1}{v}}} \cdot \Gamma\left(\frac{w+1}{v}\right),$$

$$\Re(u) > 0, \ \Re(v) > 0, \ \Re(w) > -1$$
(A.6.2)

$$K_0(x) = \frac{i\pi}{2} H_0^{(1)}(ix)$$
(A.6.3)

$$\int_0^\infty \xi^\mu K_0(b\xi) \, d\xi = 2^{\mu-1} b^{-\mu-1} \Gamma^2\left(\frac{1+\mu}{2}\right)$$
(A.6.4)

$$\Gamma(1+nH)\,\Gamma(1-nH) = \frac{\pi nH}{\sin(\pi nH)}$$
(A.6.5)

$$\int_0^{2\pi} \exp[-i(a\cos\varphi + b\sin\varphi)] \, d\varphi = J_0\left(\sqrt{a^2+b^2}\right)$$
(A.6.6)

$$\int_0^\infty J_0(b\tau) \exp(-u\tau)\tau\, d\tau = \frac{u}{\left(u^2 + b^2\right)^{3/2}} \tag{A.6.7}$$

$$\int_0^\infty J_0(b\tau) \exp\left(-u\tau^2\right)\tau\, d\tau = \frac{1}{2u} \exp\left(-\frac{b^2}{4u}\right) \tag{A.6.8}$$

Chapter 7

$$\nabla (\mathbf{A} \cdot \mathbf{B}) = (\mathbf{A} \cdot \nabla)\mathbf{B} + (\mathbf{B} \cdot \nabla)\mathbf{A} + \mathbf{A} \times (\nabla \times \mathbf{B}) + \mathbf{B} \times (\nabla \times \mathbf{A}) \tag{A.7.1}$$

$$\exp\{ia \sin \xi\} = \sum_{m=-\infty}^{+\infty} J_m(a) \exp\{im\xi\} \tag{A.7.2}$$

$$J_m(-a) = (-1)^m J_m(a) \tag{A.7.3}$$

$$\prod_{p=0}^{P-1} \sum_{m=-\infty}^{+\infty} J_m(a_p) \exp\{im\, \xi_p\}$$

$$= \sum_{m_0=-\infty}^{+\infty} \cdots \sum_{m_{P-1}=-\infty}^{+\infty} \left[\exp\left(i \sum_{p=0}^{P-1} m_p \xi_p\right) \prod_{p=0}^{P-1} J_{m_p}(a_p) \right] \tag{A.7.4}$$

Appendix B
Glossary

a	surface tangential field
α	spectral slope
B(r)	magnetic induction in phasor domain
b	surface tangential field
b(r, t**)**	magnetic induction in space-time domain
$C(\mathbf{r}_1, \mathbf{r}_2)$	Autocovariance (or covariance) function
CDF	Cumulative-Distribution Function
D(r)	electric induction in phasor domain
d(r, t**)**	electric induction in space-time domain
D	fractal dimension
$\delta(x)$	Dirac function
E(r)	electric field in phasor domain
e(r, t**)**	electric field in space-time domain
$\mathbf{E}^{(i)}$	incident electric field
$\mathbf{E}^{(s)}$	scattered electric field
$E^{(s)}_{pq}$	polarized components of the scattered electric field
EBCM	Extended-Boundary-Condition Method
ε	permittivity

273

f	electromagnetic frequency
$F(\zeta, \mathbf{r})$	Cumulative-Distribution Function
$F_{pq}(\cdot)$	polarization scattering coefficients
fBm	fractional Brownian motion
FT	Fourier Transform
$g(\mathbf{r})$	filter function in spatial domain
$G(\boldsymbol{\kappa})$	filter function in spectral domain
$\mathbf{G}(\mathbf{r} - \mathbf{r}')$	dyadic Green's function in spatial domain
GO	Geometric Optics
$\Gamma(x)$	Gamma function
$\hat{\mathbf{h}}$	perpendicular polarization unit vector
$\mathbf{H}(\mathbf{r})$	magnetic field in phasor domain
$\mathbf{h}(\mathbf{r}, t)$	magnetic field in space-time domain
$H_m^{(1)}(x)$	Hankel function of first kind and order m
H	Hurst coefficient (or exponent)
HB	Hausdorff-Besicovitch dimension
$\boldsymbol{I}$	unitary dyadic
$\Im$	imaginary part
IEM	Integral-Equation Method
$I_D^{\pm}(\mathbf{r})$	Dirichlet-type surface scalar integral
$I_N^{\pm}(\mathbf{r})$	Neumann-type surface scalar integral
$\mathbf{J}(\mathbf{r})$	current density in phasor domain
$\mathbf{j}(\mathbf{r}, t)$	current density in space-time domain
$J_m(x)$	Bessel function of first kind and order m
$\mathbf{k} = \mathbf{k}_\perp + k_z \hat{\mathbf{z}}$	vector wavenumber
K	radius of curvature
κ_e	effective spectral bandwidth
κ_0	fundamental-tone wavenumber, WM function
$K_m(x)$	Kelvin function of order m
KA	Kirchhoff approximation
L	correlation length
L_e	effective correlation length

Appendix B ◊ Glossary

λ	electromagnetic wavelength
μ	permittivity
$\mu(\mathbf{r})$	statistical mean
$N(0, \sigma^2)$	zero mean σ^2 variance Gaussian distribution
NRCS	Normalized Radar Cross-Section
ν	tone wavenumber spacing coefficient, WM function
ω	electromagnetic angular frequency
Ω	incoherency parameter
$\hat{\mathbf{p}}$	parallel polarization unit vector
$p(\zeta, \mathbf{r})$	probability-density function
$p(\zeta_1, \zeta_2; \mathbf{r}_1, \mathbf{r}_2)$	second order probability-density function
pdf	probability-density function
PDS	Power-Density Spectrum
PO	Physical Optics
$Q(\tau)$	structure function
$\mathbf{r} = \mathbf{r}_\perp + z\hat{\mathbf{z}}$	space vector coordinate
$\Re$	real part
R_h	vertical polarization Fresnel reflection coefficient
R_v	horizontal polarization Fresnel reflection coefficient
$R(\mathbf{r}_1, \mathbf{r}_2)$	autocorrelation function
rect$[x/X]$	unitary function of spatial width X
RCS	Radar Cross Section
$\rho(\tau)$	correlation coefficient
s	surface incremental standard deviation
s^2	surface incremental variance
S	scattering matrix
S_0	spectral amplitude
SPM	small-perturbation method
SSS	Strict-Sense Stationary
σ	standard deviation
$\sigma^2(\mathrm{r})$	variance

σ'^2	slope variance
σ''^2	radius of curvature variance
t	time
T	topothesy
$\boldsymbol{\tau} = \mathbf{r}_1 - \mathbf{r}_2$	vector distance
ζ	intrinsic impedance of space
$W(\kappa)$	power spectrum
WM	Weierstrass-Mandelbrot
WSS	Wide-Sense Stationary
1-D	one-dimensional
2-D	two-dimensional
3-D	three-dimensional
$\langle x \rangle$	statistical mean of x
$\lceil x \rceil$	upper integer ceiling of x
$\otimes$	convolution operator
$\triangleq$	defined as

Appendix C

References

Abramowitz, M., and I. A. Stegun. 1970. *Handbook of Mathematical Functions.* New York: Dover Publications.

Agnon, Y., and M. Stiassnie. 1991. "Remote Sensing of the Roughness of a Fractal Sea Surface." *J Geophys Res* 96 (C7): 12773–12779.

Austin, T. R., A. W. England, and G. H. Wakefield. 1994. "Special Problems in the Estimation of Power-Law Spectra as Applied to Topographical Modeling." *IEEE Trans Geosci Remote Sensing* 32 (4): 928–939.

Bahar, E., D. E. Barrick, and M. A. Fitzwater. 1983. "Computations of Scattering Cross Sections for Composite Surfaces and the Specification of the Wavenumber Where Spectral Splitting Occurs." *IEEE Trans Antennas Propag* 31 (5): 698–709.

Beckmann, P., and A. Spizzichino. 1987. *The Scattering of Electromagnetic Waves from Rough Surfaces.* Norwood, MA: Artech House.

Berry, M. V., and T. M. Blackwell. 1981. "Diffractal Echoes." *J Phys A Math Gen* 14:3101–3110.

Berry, M. V., and Z. V. Lewis. 1980. "On the Weierstrass-Mandelbrot Fractal Function." *Proc R Soc London, Sect A* 370:459–484.

Besicovitch, A. S. 1932. *Almost Periodic Functions.* London: Cambridge University Press.

Brown, S. R., and C. H. Scholz. 1985. "Broad Band Study of the Topography of Natural Rock Surfaces." *J Geophys Res* 90 (B14): 12575–12582.

Carlson, A. B. 1986. *Communication Systems.* New York: McGraw-Hill.

Chen, J., T. K. L. Lo, H. Leung, and J. Litva. 1996. "The Use of Fractals for Modeling EM Waves Scattering from Rough Sea Surface." *IEEE Trans Geosci Remote Sensing* 34 (4): 966–972.

Chew, M. F., and A. K. Fung. 1988. "A Numerical Study of the Regions of Validity of the Kirchhoff and Small-Perturbation Rough Surface Scattering Models." *Radio Sci* 23 (March–April): 163–170.

Chew, W. C. 1995. *Waves and Fields in Inhomogeneous Media.* New York: IEEE Press.

Chou, H. T., and J. T. Johnson. 1998. "A Novel Acceleration Algorithm for the Computation of Scattering from Rough Surfaces with the Forward-Backward Method." *Radio Sci* 33:1277–1287.

Chuang, S. L., and J. A. Kong. 1981. "Scattering of Waves from Periodic Surfaces." *Proc IEEE* 69 (9): 1132–1154.

Collin, R. E. 1960. *Field Theory of Guided Waves.* New York: McGraw-Hill.

Evans, D. L., T. G. Farr, and J. J. van Zyl. 1992. "Estimates of Surface Roughness Derived from Synthetic Aperture Radar (SAR) Data." *IEEE Trans Geosci Remote Sensing* GE-30:382–389.

Falconer, K. 1990. *Fractal Geometry.* Chichester, England: John Wiley.

Feder, J. S. 1988. *Fractals.* New York: Plenum Press.

Flandrin, P. 1989. "On the Spectrum of Fractional Brownian Motions." *IEEE Trans Inf Theory* 35:197–199.

Franceschetti, G. 1997. *Electromagnetics. Theory, Techniques, and Engineering Paradigms.* New York: Plenum Press.

Fung, A. K. 1994. *Microwave Scattering and Emission: Models and Their Applications.* Norwood, MA: Artech House.

Fung, A. K., Z. Li, and K. S. Chen. 1992. "Backscattering from a Randomly Rough Dielectric Surface." *IEEE Trans Geosci Remote Sensing* 30(2): 356–369.

Glazman, R. E. 1990. "Near-Nadir Radar Backscatter from a Well Developed Sea." *Radio Sci* 25:1211–1219.

Gradshteyn, I. S., and I. M. Ryzhik. 1980. *Table of Integrals, Series and Products.* New York: Academic Press.

Huang, J., and D. L. Turcotte. 1990. "Fractal Image Analysis: Application to the Topography of Oregon and Synthetic Images." *J Opt Soc Am A Opt Image Sci* 7:1124–1130.

Ishimaru, A. 1993. *Wave Propagation and Scattering in Random Media.* New York: Academic Press.

Jaggard, D. L. 1990. "On Fractal Electrodynamics." In *Recent Advances in Electromagnetic Theory*, ed. H. N. Kritikos and D. L. Jaggard, 183–223. Berlin: Springer-Verlag.

Jaggard, D. L., and X. Sun. 1990a. "Fractal Surface Scattering: A Generalized Rayleigh Solution." *J Appl Phys* 68:5456–5462.

———. 1990b. "Scattering from Fractally Corrugated Surfaces." *J Opt Soc Am A* 7:1131–1139.

Jakeman, E., and P. N. Pusey. 1976. "A Model for Non-Rayleigh Sea-Echo." *IEEE Trans Antennas Propag* AP-24:806–814.

Jao, J. K. 1984. "Amplitude Distribution of Composite Terrain Radar Clutter and the K-Distribution." *IEEE Trans Antennas Propag* AP-32:1049–1061.

Jao, J. K., and M. Elbaum. 1978. "First-Order Statistics of a Non-Rayleigh Fading Signal and Its Detection." *Proc IEEE* 7:781–789.

Kong, J. A. 1986. *Electromagnetic Wave Theory.* New York: John Wiley.

Lo, T., H. Leung, J. Litva, and S. Haykin. 1993. "Fractal Characterisation of Sea-Scattered Signals and Detection of Sea-Surface Targets." *IEE Proc F* 140:243–250.

Mandelbrot, B. B. 1983. *The Fractal Geometry of Nature.* New York: W. H. Freeman.

Mickelson, A. R., and D. L. Jaggard. 1979. "Electromagnetic Wave Propagation in Almost Periodic Media." *IEEE Trans Antennas Propag* A-P27 (1): 34–40.

Oh, Y., K. Sarabandi, and F. T. Ulaby. 1992. "An Empirical Model and an Inversion Technique for Radar Scattering from Bare Soil Surfaces." *IEEE Trans Geosci Remote Sensing* 30 (2): 370–381.

Oliver, C. J. 1984. "A Model for Non-Rayleigh Scattering Statistics." *Opt Acta* 31:701–722.

Ouchi, K., S. Tajbakhsh, and R. E. Burge. 1987. "Dependence of Speckle Statistics on Backscatter Cross-section Fluctuations in Synthetic Aperture Radar Images of Rough Surfaces." *IEEE Trans Geosci Remote Sensing* GE-25:623–628.

Papoulis, A. 1965. *Probability, Random Variables and Stochastic Processes.* New York: McGraw-Hill.

Prudnikov, A. P., Y. A. Brychkov, and O. I. Marichev. 1990. *Integrals and Series.* New York: Gordon & Breach.

Rodriguez, E. 1989. "Beyond the Kirchhoff Approximation." *Radio Sci* 24:681–693.

Savaidis, S., P. Frangos, D. L. Jaggard, and K. Hizanidis. 1997. "Scattering from Fractally Corrugated Surfaces with Use of the Extended Boundary Condition Method." *J Opt Soc Am* 14 (2): 475–485.

Stewart, C. V., B. Moghaddam, K. J. Hintz, and L. M. Novak. 1993. "Fractional Brownian Motion Models for Synthetic Aperture Radar Imagery Scene Segmentation." *Proc IEEE* 81:1511–1522.

Tsang, L., J. A. Kong, and R. T. Shin. 1985. *Theory of Microwave Remote Sensing*. New York: John Wiley.

Tsang, L., J. A. Kong, and K. H. Ding. 2000, *Scattering of Electromagnetic Waves, Vol. 1:Theory and Applications*, New York: Wiley Interscience.

Tsang, L., J. A. Kong, K. H. Ding and C. O. Ao. 2001. *Scattering of Electromagnetic Waves, Vol. 2: Numerical Simulations*, New York: Wiley Interscience.

Tsang, L., and J. A. Kong. 2001. *Scattering of Electromagnetic Waves, Vol. 3: Advanced Topics*, New York: Wiley Interscience.

Ulaby, F. T., and M. C. Dobson. 1989. *Handbook of Radar Scattering Statistics for Terrain*. Norwood, MA: Artech House.

Ulaby, F. T., R. K. Moore, and A. K. Fung. 1982. *Microwave Remote Sensing*, Reading, MA: Addison-Wesley.

Voss, R. F. 1985. "Random Fractal Forgeries." In *Fundamental Algorithms for Computer Graphics*, ed. R. A. Earnshaw, 805–835. New York: Springer-Verlag.

Waterman, P. C. 1975. "Scattering by Periodic Surfaces." *J Acoust Soc Amer* 57:791–802.

West, J. C., B. S. O'Leary, and J. Klinke. 1998. "Numerical Calculation of Electromagnetic Scattering from Measured Wind-Roughened Water Surfaces." *Int J Remote Sensing* 19 (7): 1377–1393.

Wornell, G. W. 1995. *Signal Processing with Fractals: A Wavelet Based Approach*. Upper Saddle River, NJ: Prentice Hall.

Yocoya, N., K. Yamamoto, and N. Funakubo. 1989. "Fractal-based Analysis and Interpolation of 3D Natural Surface Shapes and Their Application to Terrain Modeling." *Computer Vision, Graphics, and Image Processing* 46:284–302.

References Involving the Authors of This Book

Ceraldi, E., G. Franceschetti, A. Iodice, D. Riccio, G. Ruello. 2005. "Estimating the Soil Dielectric Constant via Scattering Measurements along the Specular Direction", *IEEE Trans Geosci Remote Sensing*, GE-43 (February): 295–305.

Collaro, A., G. Franceschetti, M. Migliaccio, and D. Riccio. 1999. "Gaussian Rough Surfaces Scattering and Kirchhoff Approximation." *IEEE Trans Antennas Propag* AP-47 (February): 392–398.

Coltelli, M., L. Dutra, G. Fornaro, G. Franceschetti, R. Lanari, M. Migliaccio, J. R. Moreira, K. P. Papathanassiou, G. Puglisi, D. Riccio, and M. Schwäbisch. 1996. "SIR-C/X-SAR Interferometry Over Mt. Etna: DEM Generation, Accuracy Assessment and Data Interpretation." *Deutsche Forschungsanstalt fur Luft-und Raumfahrt Book*, Munich (Germania), ISSN 0939-2963, pp.91, 1996.

Coltelli, M., G. Fornaro, G. Franceschetti, R. Lanari, M. Migliaccio, J. R. Moreira, K. P. Papathanassiou, G. Puglisi, D. Riccio, and M. Schwäbisch. 1996. "SIR-C/X-SAR Multifrequency Multipass Interferometry: A New Tool for Geological Interpretation." *J Geophys Res* 101-E10 (October): 23127–23148.

Franceschetti, G., P. Callahan, A. Iodice, D. Riccio, S. Wall. 2006. "Titan, Fractals and Filtering of Cassini Altimeter Data", *IEEE Trans Geosci Remote Sensing*, GE-44 (August): 2055–2062.

Franceschetti, G., A. Iodice, S. Maddaluno, and D. Riccio. 2000, "A Fractal Based Theoretical Framework for the Retrieval of Surface Parameters from Electromagnetic Backscattering Data." *IEEE Trans Geosci Remote Sensing* GE-38 (March): 641–650.

Franceschetti, G., A. Iodice, M. Migliaccio, and D. Riccio. 1997. "The Effect of Surface Scattering on the IFSAR Baseline Decorrelation." *J Electromagn Waves Appl* 11 (March): 353–370.

———. 1999a. "Fractals and the Small Perturbation Scattering Model." *Radio Sci* 34-5 (September–October): 1043–1054.

———. 1999b. "Scattering from a Fractal Sea Surface." *Atti della Fondazione Giorgio Ronchi* LIV 3-4 (May–August): 455–462.

———. 1999c. "Scattering from Natural Rough Surfaces Modelled by Fractional Brownian Motion Two-Dimensional Processes." *IEEE Trans Antennas Propag* AP-47 (September): 1405–1415.

Franceschetti, G., A. Iodice, and D. Riccio. 2000. "Scattering from Dielectric Random Fractal Surfaces via Method of Moments." *IEEE Trans Geosci Remote Sensing* GE-38 (July): 1644–1655.

———. 2001. "Scattering from Natural Surfaces." In *Scattering*, R. Pike, P. Sabatier, 467–485. London: Academic Press.

Franceschetti, G., A. Iodice, D. Riccio, and G. Ruello. 2001. "Electromagnetic Scattering from Fractal Profiles: The Extended Boundary Condition Method." *Atti della Fondazione Ronchi* LVI (August–October): 1005–1012.

———. 2002. "Fractal Surfaces and the Extended Boundary Conditions." *IEEE Trans Geosci Remote Sensing* GE-40 (May): 1918–1931.

Franceschetti, G., A. Iodice, D. Riccio, G. Ruello. 2005. "Extended Boundary Condition Method for Scattering and Emission from Natural Surfaces Modeled by Fractals", *IEEE Trans Geosci Remote Sensing*, GE-43 (May): 1115–1125.

Franceschetti, G., D. Riccio. 2004. "Stochastic Surface Models for Electromagnetic in *Electromagnetics in a Complex World: Challenges and Perspectives*, I. M. Pinto, V. Galdi, L. B. Felsen, eds, 207–214. New-York: Springer-Verlag.

Ruello, G., P. Blanco, A. Iodice, J. J. Mallorqui, D. Riccio, A. Broquetas, G. Franceschetti. 2006. "Synthesis, Construction and Validation of a Fractal Surface",*IEEE Trans Geosci Remote Sensing*, GE-44 (June): 1403–1412.

Franceschetti, G., M. Migliaccio, and D. Riccio. 1996. "An Electromagnetic Fractal-Based Model for the Study of the Fading." *Radio Sci* 31-6 (November–December): 1749–1759.